"The authors are perceptive and their generalities are relevant to subjective experiences and behavior of individuals....The book is very readable and of particular interest to those who are concerned with the difference between the psychotherapies of psychotics and neurotics." —*Psychiatric Quarterly*

"For those who would like data or science, this book will be disappointing; for those who prefer humanism this book will be exciting."

—*Choice*

"These papers are sufficiently interesting in themselves, but Barry Stevens' perceptive and sensitive reactions to each author makes the book an even more exciting encounter. We are encouraged to let others be 'themselves,' to see more readily what we are ourselves, to drop our fears of what others think, and learn to be free. Other therapies tend to make man predictable but, as is pointed out in the book, 'the goals of a humanistic psychotherapy are neither predictability nor control. In fact, the more successful psychotherapy is, the less predictable the individual becomes....' Stevens chooses creativity as a fundamental value in considering the nature of man."

—*Canada's Mental Health*

Person to Person:

The Problem of Being Human

A New Trend in Psychology

by

Carl R. Rogers and Barry Stevens

with contributions from:

Eugene T. Gendlin

John M. Shlien

Wilson Van Dusen

PUBLISHED BY POCKET BOOKS NEW YORK

And one clock stopped—
and knew the meaning of time.

Contents

Foreword

Carl R. Rogers, Ph.D.

This is an unusual book, put together by a unique person, and built around some papers which present an all too uncommon point of view. Let me explain.

In the first place each paper—some of my own, others by Drs. Gendlin, Shlien, and Van Dusen—is set in a context of warm human reaction to the paper. Barry Stevens regards each of these papers as a sort of oasis in current professional literature, and has lovingly placed each of them in a setting composed of her own very personal associations to the theme of the writer. Her comments are not comments on the paper. They are not a review of the paper. They are the highly personal feelings and thoughts which the paper triggered off in her. It is as though a friend told you of many responses set off in himself by something he had read. So you are stimulated to read the material yourself to see what *you* can get out of it. This seems like a natural approach, but it is certainly not a conventional one. It is simply *not* the way books are written or compiled.

But this is not so surprising when we consider Barry Stevens. She is not a person easily categorized. Though she knows, and is in correspondence with, many of the great and near-great in our Western culture she has no position, no status, no professional classification except the vague term, "writer." (I think she would prefer the term "amateur," because she deeply believes that the "amateur" and the "professional" complement each other.) She is independent in her thought and in her life, continually striving to break out of the bonds which tend to hold all of us. She wrote much of the material for this book while a guest of the Western Behavioral Sciences Institute, and in her quiet, effortless way became, during that period, an important figure in the lives of many of the individuals who work there. She has somehow achieved in her life a *wisdom* which seems all too rare in these days when *knowledge* has become so all-important. Many readers may find her personal statements more valuable and more rewarding than the writings they were intended to frame. She often sees through to the heart of the matter in a most perceptive way. But I believe the uniqueness of her person shows through in her writing, so I will let the reader discover her for himself.

The book is built around seven papers which start from an assumption unusual in psychology today. The assumption is that the subjective human being has an importance and a value which is basic: that no matter how he may be labeled or evaluated he is a human person first of all, and most deeply. He is not only a machine, not only a collection of stimulus-response bonds, not an object, not a pawn. So while the papers cover a variety of topics, and in at least three instances deal with individuals labeled "abnormal," they are all basically about persons.

For this reason it is felt that they may appeal to persons—persons who, like the authors, are seeking for a way in which life can best be chosen, and most rewardingly lived. It will be obvious that the papers appealed to one person, Barry Stevens, and her reactions—sometimes charming, sometimes moving, sometimes critical, sometimes profound—form the links between the themes of the authors.

This book would not have been written, compiled, or published, if it had not been the belief of all concerned that it has a relevance for human beings today. It seems possible to us that the reader may come to feel something like this: "This book helps me to understand myself a little better, so now I understand the other a bit better, and to this degree I am a little less baffled by both of us." Or to put it in a slightly different fashion, it is our hope that the person who reads it will find in it permission to be and to become more of himself.

Introduction

Barry Stevens

This book is intended for anyone who is interested in it; anyone who is not interested should not read it. That is a simple statement of what this book is chiefly about: the importance of choosing for ourselves, regardless of what anyone else tells us is good, or bad. My view of why this is not easy to do is summed up in the *Curtain Raiser.* The one person the world seems hell-bent on my not living with is *me.*

I write of my own experience because I am the only person I can really know the happenings in, but I write of these happenings not as unique, but as something that it seems to me is true of all of us. Even when my experience is not usual in the external events, it seems to me that in some way or to some degree you must know through your experience the inner happenings that I write about. Out of my knowing of an uncountable number of other people. I have a very strong feeling that I am not exceptional. I like this. I wish that I had been aware of it all my life. I write about myself, but the comments that please me say, "That's me you're writing about," and "It says something to me. It sounds like it should have been me saying it, as though you were me when you wrote it."

We are pushed into a lot of phoney "togetherness" that I resent. But there is a real being-together that no one has to be pushed into because it springs immediately from ourselves.

In this book there are several papers about psychotherapy with schizophrenics which include the therapists themselves—what goes on in them, what goes on between them and someone else, their approach to being with another person. When I first read them, I had a deep feeling that this is basically the way that all of us should approach each other instead of the way that we usually do. I gave these papers to some other people to read without mentioning my own response. I quote in full what they said when they gave the papers back to me:

> "It's what we should be doing." (A Swedish graduate student in English literature at one of our universities.)
>
> "It's what we should be doing—with our family, and our friends." (A middle-aged woman working in public health.)
>
> "It's so awful that it needs to be said, because it's what all of us should be doing all the time." (A young cellist who had never gone to college.)

Not one of them mentioned "schizophrenic." That was not in my mind, either. It was just people to me, and what I would like to receive from others and be able to give to them. I had found this way of being with people easy with most Hawaiians, Orientals in Hawaii, Navajos, and Hopis, because that was their approach to me, but in my own society I find it difficult, except occasionally with some people. I was very cross with Carl Rogers when he wrote that the hours he spent in therapy were the most exciting to him, because that is what I want to live with all the time and why should a person have to go to a therapist to have it? Besides, it seemed to me that lack of it is what disorganizes many people so much that they have to be in mental hospitals. Why shouldn't we have this kind of being with people everywhere, so they wouldn't have to go to therapists or be in hospitals?

This is my view, which has led me to put this book together. The professional papers in it present two approaches: one is the direct interaction between people; the other is the knowledge *about* people that has been arrived at through psychotherapy. In myself, I have found that these two approaches complement each other, and that I do better with both than with either one alone. In my early days of living with people in other cultures I just enjoyed living with them in this respect, without trying to undersand it. (I write "in this respect" because there are ways in which I could not live fully with the Hopis or Navajos—beliefs and customs which

I could not accept—but in *this* way I enjoyed being with them very much and the differences in beliefs or customs made no difference, either to them or to me.) When I was away from them and back in my own culture, I was aware of something that I had lost but I did not understand this. I could fall into their ways with them, but I couldn't reproduce it anywhere else. I couldn't talk about it, either. More recently when I have been with them—the most recently for five months—I have noticed and tried to understand what it is that I like, and I find that this helps very much in making it possible for me to *be,* in their way, when I am away from them—at least sometimes. And even when I fail, I am annoyed, but not lost—confused and bewildered—because of what I understand. It seems to me that the opposite may be what happens to Navajos and Hopis and Hawaiians when they get swamped by our way of doing; they lose something without knowing what they have lost, because they do not understand it. That happened to me too within my own society. I started out very fortunately. My life went well. Other people remarked on this. I was not afraid of losing what I had because it was myself, and how could you lose yourself? But I lost it. Now, I am getting it back at least somewhat through understanding, which, if I had had it in the first place, would have made it possible for me not to lose what I had.

But understanding alone is not enough. When I understand something and do not put this into action, nothing has been accomplished either in the outside world or within myself. A Boston psychiatrist calls it "the kiss of death" when a patient says, "I understand!" I know what he means because when I was being my own therapist I hit on something that was really blocking me and kept telling myself that as long as this was clear to me, as long as I understood how it had come about, I didn't need to *do* anything about it. "It's all right." It wasn't. Nothing *happened* until I acted on what I knew. It seems to be one of the follies of my intellect that it can think that what I know, I am, and that words take care of everything.

Carl Rogers and I were born in the same year. Other than that, there is very little in common in our external lives. But in our internal lives, he went very much his own way and I went very much my own way, and it was through going our *own* ways that, a few years ago, we got together through correspondence, and then last year we met. When this happened, we found ourselves so much in agreement about life and people that it was like matching. It seems to me that this is what happens when each man "steps to the music that he hears"—his own inner trumpet. I certainly didn't hear mine all the time. Twice I got into such mental chaos that it seemed I'd never find my way out. Both times, I got out of it by listening to myself again.

When I first thought of this book, I expected that it would be the usual presentation of papers with some comments by the editor—myself. But then Carl Rogers wrote that he hoped there would be more of myself in the book than that. I had found that in talking with people about his papers, often a description of something from my own life helped them to latch onto what he meant. This seemed to come about through a kind of mutuality of experience: although my own experience was different from theirs in details of setting, persons, incidents, at the same time something about it brought into their consciousness an experience of their own which made the meaning clear within themselves. I hope that my writing in this book may do the same for you whom I do not know.

If this bringing *your* experience into awareness does not happen, then my work is useless. If anything that I have written is accepted without this awareness, then it is harmful because, in my view, that is how the trouble comes about: for various reasons we take in what has no connection with ourselves and think of it *as* ourselves. This happened so much with me through my husband that when I realized it, after his suicide, I felt that I could not let a man love me because how could he or I know whether what he loved in me was my husband or myself?

On the other hand, if I have lived for weeks or months with something that you have known only for a moment, that moment is real *to you,* and if it is brought into your awareness that is good. *My* experience then has been used to fish-hook yours into consciousness, and the fishhook can be thrown away.

For this to happen, it seems necessary to read with the mind turned inward, seeking "Where are the connections with myself?" Time is necessary for this—time for a dialogue between the reader and the book. I find that reading a paragraph or a page, when that is all I have time for, is more valuable to me than rushing through a lot of pages when I have an hour. The dialogue betwen that passage and myself seems to go on underground while I am doing something else. I notice this when it surfaces for a moment, showing me what is going on.

Just recently, I read a book by going straight through it from front to back, but skipping the sections that didn't interest me. When I had got to the end, I was interested in going back and reading the parts that I had skipped. Then I read again the parts that I had read the first time.

These are only two of the ways that I have read books, and I don't recommend that either of them be followed. The important thing for me to notice is, "Am I bored?" If so, then either I stop reading or flip the pages, in any old fashion, until something in the book and I come easily together and I read because I am interested. Something works itself out between *this* book and me by playing around with it.

One of the problems to me in my writing for this book was, "How far shall I directly relate my own writing to the papers?" I asked some people about this. A young man said, "I like to make my own connections. I feel *good* when they pop into my head." Then he said, with equal vigor, "But when I don't see them, then I want them to be made for me." I think that says it. There is no answer to the problem. The professional papers are here for anyone to interact with on his own, and if my writing is annoying, it can be skipped.

One paper in this book has not been previously published. The others have been published in professional journals and, in two instances, in professional books. They are not easy for most of us to find. My son found one of them in a waste basket or we might never have known of it at all. This book at least makes them available to all of us, and to me it is to all of us they speak.

As I see it, the client-centered approach to interpersonal relations cannot be *followed* in the usual sense, which is a kind of copying. It requires that a person start from the same point that Rogers does—a point that has to be arrived at by each person in himself. From that point which is central to himself, of necessity his mode will be his own, his course an exploratory one, made with the knowledge that there are no final answers, that a readiness to correct mistakes works out better than trying to make them. The papers by Eugene Gendlin and John Shlien show that they genuinely use the client-centered approach, and their own individuality is apparent. In the same way, Wilson Van Dusen writes, "I see considerable similarity between myself and Carl Rogers. The difference would be in style and in the way we would describe things. This matter of style is something which belies any attempt at the one proper best psychotherapy. What is proper for one, is not for another."

It seems to me that this is true of all of us in our relations with each other, whether as parent, spouse, friend, acquaintance or person simply met and lived with briefly: we need to start from the same base, which we each have to find in ourselves, but after that "what is proper for one, is not for another."

At this point, I suppose that I should make some acknowledgments, but when I read acknowledgments they always seem so meaningless, while what I feel for those who made this book possible is deeply a part of my living and my life. I think they know this. So let it stay there, person to person.

Person to Person

Curtain Raiser

Barry Stevens

In the beginning, I was one person, knowing nothing but my own experience.

Then I was told things, and I became two people: the little girl who said how terrible it was that the boys had a fire going in the lot next door where they were roasting apples (which was what the women said)—and the little girl who, when the boys were called by their mothers to go to the store, ran out and tended the fire and the apples because she loved doing it.

So then there were two of I.

One I always doing something that the other I disapproved of. Or other I said what I disapproved of. All this argument in me so much.

In the beginning was I, and I was good.

Then came in other I. Outside authority. This was confusing. And then other I became *very* confused because there were so many different outside authorities.

Sit nicely. Leave the room to blow your nose. Don't do that, that's silly. Why, the poor child doesn't even know how to pick a bone! Flush the toilet at night because if you don't it makes it harder to clean. DON'T FLUSH THE TOILET AT NIGHT—you wake people up! Always be nice to people. Even if you don't like them, you mustn't hurt their feelings. Be frank and honest. If you don't tell people what you think of them, that's cowardly. Butter knives. It is important to use butter knives. Butter knives? What foolishness! Speak nicely. Sissy! Kipling is wonderful! Ugh! Kipling (turning away)

The most important thing is to have a career. The most important thing is to get married. The hell with everyone. Be nice to everyone. The most important thing is sex. The most important thing is to have money in the bank. The most important thing is to have everyone like you. The most important thing is to dress well. The most important thing is to be sophisticated and say what you don't mean and don't let anyone know what you feel. The most important thing is to be ahead of everyone else. The most important thing is a black seal coat and china and silver. The most important thing is to be clean. The most important thing is to always pay your debts. The most important thing is not to be taken in by anyone else. The most important thing is to love your parents. The most important thing is to work. The most important thing is to be independent. The most important thing is to speak correct English. The most important thing is to be dutiful to your husband. The most important thing is to see that your children behave well. The most important thing is to go to the right plays and read the right books. The most important thing is to do what others say. And others say all these things.

All the time, *I* is saying, live with life. That is what is important.

But when I lives with life, other I says no, that's bad. All the different other I's say this. It's dangerous. It isn't practical. You'll come to a bad end. Of course . . . everyone felt that way once, the way you do, but *you'll learn!*

Out of all the other I's some are chosen as a pattern that is me. But there are all the other possibilities of patterns within what all the others say which come into me and become other I which is not myself, and sometimes these take over. Then who am I?

I does not bother about who am I. I *is,* and is happy being. But when I is happy being, other I says get to work, do something, do something worthwhile. I is happy doing dishes. "You're weird!" I is happy being with people saying nothing. Other I says talk. Talk, talk, talk. I gets lost.

I knows that things are to be played with, not possessed. I likes putting things together, lightly. Taking things apart, lightly. "You'll never have anything!" Making things of things in a way that the things themselves take part in, putting themselves together with surprise and delight to I. "There's no money in that!"

I is human. If someone needs I gives. "You can't do that! You'll never have anything for yourself! We'll have to support you!"

I loves. I loves in a way that other I does not know. I loves. "That's too warm for friends!" "That's too cool for lovers!" "Don't feel so bad, he's just a friend. It's not as though you loved him." "How can you let him go? I thought you loved him?" So cool the warm for friends and hot up the love for lovers, and I gets lost.

So both I's have a house and a husband and children and all that, and friends and respectability and all that, and security and all that, but both I's are confused because other I says, "You see? You're lucky," while I goes on crying. "What are you crying about? Why are you so ungrateful?" I doesn't know gratitude or ingratitude, and cannot argue. I goes on crying. Other I pushes it out, says "I am happy! I am very lucky to have such a fine family and a nice house and good neighbors and lots of friends who want me to do this, do that." I is not reason-able, either. I goes on crying.

Other I gets tired, and goes on smiling, because that is the thing to do. Smile, and you will be rewarded. Like the seal who gets tossed a piece of fish. Be nice to everyone and you will be rewarded. People will be nice to you, and you can be happy with that. You know they like you. Like a dog who gets patted on the head for good behavior. Tell funny stories. Be gay. Smile, smile, smile. . . . I is crying. . . . "Don't be sorry for yourself! Go out and do things for people!" "Go out and be with people!" I is still crying, but now, that is not heard and felt so much.

Suddenly: "What am I doing?" "Am I to go through life playing the clown?" "What am I doing, going to parties that I do not enjoy?" "What am I doing, being with people who bore me?" "Why am I so hollow and the hollowness filled with emptiness?" A shell. How has this shell grown around me? Why am I proud of my children and unhappy about their lives which are not good enough? Why am I disappointed? Why do I feel so much waste?

I comes through, a little. In moments. And gets pushed back by other I.

I refuses to play the clown any more. Which I is that? "She used to be fun, but now she thinks too much about herself." I lets friends drop away. Which I is that? "She's being too much by herself. That's bad. She's losing her mind." Which mind?

*Toward a Modern Approach to Values: The Valuing Process in the Mature Person**

Carl R. Rogers

There is a great deal of concern today with the problem of values. Youth, in almost every country, is deeply uncertain of its value orientation; the values associated with various religions have lost much of their influence; sophisticated individuals in every culture seem unsure and troubled as to the goals they hold in esteem. The reasons are not far to seek. The world culture, in all its aspects, seems increasingly scientific and relativistic, and the rigid, absolute views on values which come to us from the past appear anachronistic. Even more important perhaps, is the fact that the modern individual is assailed from every angle by divergent and contradictory value claims. It is no longer possible, as it was in the not too distant historical past, to settle comfortably into the value system of one's forebears or one's community and live out one's life without ever examining the nature and the assumptions of that system.

In this situation it is not surprising that value orientations from the past appear to be in a state of disintegration or collapse. Men question whether there are, or can be, any universal values. It is often felt that we may have lost, in our modern world, all possibility of any general or cross-cultural basis for values. One natural result of this uncertainty and confusion is that there is an increasing concern about, interest in, and a searching for, a sound or meaningful value approach which can hold its own in today's world. I

* *Journal of Abnormal and Social Psychology* 1964, Vol. 68, No. 2, 160-167.

share this general concern. I have also experienced the more specific value issues which arise in my own field, psychotherapy. The client's feelings and convictions about values frequently change during therapy. How can he or we know whether they have changed in a sound direction? Or does he simply, as some claim, take over the value system of his therapist? Is psychotherapy simply a device whereby the unacknowledged and unexamined values of the therapist are unknowingly transmitted to an unsuspecting client? Or should this transmission of values be the therapist's openly held purpose? Should he become the modern priest, upholding and imparting a value system suitable for today? And what would such a value system be? There has been much discussion of such issues, ranging from thoughtful and empirically based presentations such as that of D. D. Glad, to more polemic statements. As is so often true, the general problem faced by the culture is painfully and specifically evident in the cultural microcosm which is called the therapeutic relationship.

I should like to attempt a modest approach to this whole problem. I have observed changes in the approach to values as the individual grows from infancy to adulthood. I observe further changes when, if he is fortunate, he continues to grow toward true psychological maturity. Many of these observations grow out of my experience as a therapist, where I have had the rich opportunity of seeing the way in which individuals move toward a richer life. From these observations I believe I see some directional threads emerging which might offer a new concept of the valuing process, more tenable in the modern world. I have made a beginning by presenting some of these ideas partially in previous writings; I would like now to voice them more clearly and more fully.

I would stress that my vantage point for making these observations is not that of the scholar or philosopher: I am speaking from my experience of the functioning human being, as I have lived with him in the intimate experence of therapy, and in other situations of growth, change, and development.

Some definitions

Before I present some of these observations, perhaps I should try to clarify what I mean by values. There are many definitions which have been used, but I have found helpful some distinctions made by Charles Morris. He points out that value is a term we employ in different ways. We use it to refer to the tendency of any living beings to show preference, in their actions, for one kind of object or objective rather than another. This preferential behavior he calls "operative values." It need not involve any cognitive or conceptual thinking. It is simply the value choice which is indicated behaviorally when the organism selects one object, rejects another. When the earthworm, placed in a simple Y maze, chooses the smooth arm of the Y, instead of the path which is paved with sandpaper, he is indicating an operative value.

A second use of the term might be called "conceived values." This is the preference of the individual for a symbolized object. Usually in such a preference there is anticipation or foresight of the outcome of behavior directed toward such a symbolized object. A choice such as "Honesty is the best policy" is such a conceived value.

A final use of the term might be called "objective value." People use the word in this way when they wish to speak of what is objectively preferable, whether or not it is in fact sensed or conceived of as desirable. What I have to say involves this last definition scarcely at all. I will be concerned with operative values and conceptualized values.

The Infant's Way of Valuing

Let me first speak about the infant. The living human being has, at the outset, a clear approach to values. He prefers some things and experiences, and rejects others. We can infer from studying his behavior that he prefers those experiences which maintain, enhance, or actualize his organism, and rejects those which do not serve this end. Watch him for a bit:

> Hunger is negatively valued. His expression of this often comes through loud and clear.
>
> Food is positively valued. But when he is satisfied, food is negatively valued, and the same milk he responded to so eagerly is now spit out, or the breast which seemed so satisfying is now rejected as he turns his head away from the nipple with an amusing facial expression of disgust and revulsion.

He values security, and the holding and caressing which seem to communicate security.

He values new experience for its own sake, and we observe this in his obvious pleasure in discovering his toes, in his searching movements, in his endless curiosity.

He shows a clear negative valuing of pain, bitter tastes, sudden loud sounds.

All of this is commonplace, but let us look at these facts in terms of what they tell us about the infant's approach to values. It is first of all a flexible, changing, valuing *process*, not a fixed system. He likes food and dislikes the same food. He values security and rest, and rejects it for new experience. What is going on seems best described as an organismic valuing process, in which each element, each moment of what he is experiencing is somehow weighed, and selected or rejected, depending on whether, at this moment, it tends to actualize the organism or not. This complicated weighing of experience is clearly an organismic, not a conscious or symbolic function. These are operative, not conceived values. But this process can none the less deal with complex value problems. I would remind you of the experiment in which young infants had spread in front of them a score or more of dishes of natural (that is, unflavored) foods. Over a period of time they clearly tended to value the foods which enhanced their own survival, growth, and development. If for a time a child gorged himself on starches, this would soon be balanced by a protein "binge." If at times he chose a diet deficient in some vitamin, he would later seek out foods rich in this very vitamin. He was utilizing the wisdom of the body in his value choices, or perhaps more accurately, the physiological wisdom of his body guided his behavioral movements, resulting in what we might think of as objectively sound value choices.

Another aspect of the infant's approach to value is that the source of locus of the evaluating process is clearly within himself. Unlike many of us, he *knows* what he likes and dislikes, and the origin of these value choices lies strictly within himself. He is the center of the valuing process, the evidence for his choices being supplied by his own senses. He is not at this point influenced by what his parents think he should prefer, or by what the church says, or by the opinion of the latest "expert" in the field, or by the persuasive talents of an advertising firm. It is from within his own experiencing that his organism is saying in non-verbal terms, "This is good for me." "That is bad for me." "I like this." "I strongly dislike that." He would laugh at our concern over values, if he could understand it. How could anyone fail to know what he liked and disliked, what was good for him and what was not?

The Change in the Valuing Process

What happens to this highly efficient, soundly based valuing process? By what sequence of events do we exchange it for the more rigid, uncertain, inefficient approach to values which characterizes most of us as adults? Let me try to state briefly one of the major ways in which I think this happens.

The infant needs love, wants it, tends to behave in ways which will bring a repetition of this wanted experience. But this brings complications. He pulls baby sister's hair, and finds it satisfying to hear her wails and protests. He then hears that he is "a naughty, bad boy," and this may be reinforced by a slap on the hand. He is cut off from affection. As this experience is repeated, and many, many others like it, he gradually learns that what "feels good" is often "bad" in the eyes of others. Then the next step occurs, in which he comes to take the same attitude toward himself which these others have taken. Now, as he pulls his sister's hair, he solemnly intones, "Bad, bad boy." He is introjecting the value judgment of another, taking it as his own. He has deserted the wisdom of his organism, giving up the locus of evaluation, and is trying to behave in terms of values set by another, in order to hold love.

Or take another example at an older level. A boy senses, though perhaps not consciously, that he is more loved and prized by his parents when he thinks of being a doctor than when he thinks of being an artist. Gradually he introjects the values attached to being a doctor. He comes to want, above all, to be a doctor. Then in college he is baffled by the fact that he repeatedly fails in chemistry, which is absolutely necessary to becoming a physician, in spite of the fact that the guidance counselor assures him he has the ability to pass the course. Only in counseling interviews does he begin to realize how completely he has lost touch with his organismic reactions, how out of touch he is with his own valuing process.

Let me give another instance from a class of mine, a group of prospective teachers. I asked them at the beginning of the course, "Please list for me the two or three values which you would most wish to pass on to the children with whom you will work." They turned in many value goals, but I was surprised by some of the items. Several listed such things as "to speak correctly," "to use good English, not to use words like ain't." Others mentioned neatness—"to do things according to instructions"; one explained her hope that "When I tell them to write their names in the upper right-hand corner with the date under it, I want them to do it *that way,* not in some other form."

I confess I was somewhat appalled that for some of these girls the most important values to be passed on to pupils were to avoid bad grammar, or meticulously to follow teacher's instructions. I felt baffled. Certainly these behaviors had not been *experienced* as the most satisfying and meaningful elements in their own lives. The listing of such values could only be accounted for by the fact that these behaviors had gained approval—and thus had been introjected as deeply important.

Perhaps these several illustrations will indicate that in an attempt to gain or hold love, approval, esteem, the individual relinquishes the locus of evaluation which was his in infancy, and places it in others. He learns to have a basic *dis*trust for his own experiencing as a guide to his behavior. He learns from others a large number of conceived values, and adopts them as his own, even though they may be widely discrepant from what he is experiencing. Because these concepts are not based on his own valuing, they tend to be fixed and rigid, rather than fluid and changing.

Some Introjected Patterns

It is in this fashion, I believe, that most of us accumulate the introjected value patterns by which we live. In this fantastically complex culture of today, the patterns we introject as desirable or undesirable come from a variety of sources and are often highly contradictory in their meanings. Let me list a few of the introjections which are commonly held.

Sexual desires and behaviors are mostly bad. The sources of this construct are many—parents, church, teachers.

Disobedience is bad. Here parents and teachers combine with the military to emphasize this concept. To obey is good. To obey without question is even better.

Making money is the highest good. The sources of this conceived value are too numerous to mention.

Learning an accumulation of scholarly facts is highly desirable.

Browsing and aimless exploratory reading for fun is undesirable. The source of these last two concepts is apt to be in school, the educational system.

Abstract art or "pop" art, or "op" art is good. Here the people we regard as sophisticated are the originators of the value.

Communism is utterly bad. Here the government is a major source.

To love thy neighbor is the highest good. This concept comes from the church, perhaps from the parents.

Cooperation and teamwork are preferable to acting alone. Here companions are an important source.

Cheating is clever and desirable. The peer group again is the origin.

Coca-Colas, chewing gum, electric refrigerators, and automobiles are all utterly desirable. This conception comes not only from advertisements, but is reinforced by people all over the world. From Jamaica to Japan, from Copenhagen to Kowloon, the "Coca-Cola culture" has come to be regarded as the acme of desirability.

This is a small and diversified sample of the myriads of conceived values which individuals often introject, and hold as their own, without ever having considered their inner organismic reactions to those patterns and objects.

Common Characteristics of Adult Valuing

I believe it will be clear from the foregoing that the usual adult—I feel I am speaking for most of us—has an approach to values which has these characteristics:

The majority of his values are introjected from other individuals or groups significant to him, but are regarded by him as his own.

The source or locus of evaluation on most matters lies outside of himself.

The criterion by which his values are set is the degree to which they will cause him to be loved or accepted.

These conceived preferences are either not related at all, or not clearly related, to his own process of experiencing.

Often there is a wide and unrecognized discrepancy between the evidence supplied by his own experience, and these conceived values.

Because these conceptions are not open to testing in experience, he must hold them in a rigid and unchanging fashion. The alternative would be a collapse of his values. Hence his values are "right"—like the law of the Medes and the Persians, which changeth not.

Because they are untestable, there is no ready way of solving contradictions. If he has taken in from the community the conception that money is the summum bonum and from the church the conception that love of one's neighbor is the highest value, he has no way of discovering which has more value for *him*. Hence a common aspect of modern life is living with absolutely contradictory values. We calmly discuss the possibility of dropping a hydrogen bomb on Russia, but then find tears in our eyes when we see headlines about the suffering of one small child.

Because he has relinquished the locus of evaluation to others, and has lost touch with his own valuing process, he feels profoundly insecure and easily threatened in his values. If some of these conceptions were destroyed, what would take their place? This threatening possibility makes him hold his value conceptions more rigidly or more confusedly, or both.

The Fundamental Discrepancy

I believe that this picture of the individual, with values mostly introjected, held as fixed concepts, rarely examined or tested, is the picture of most of us. By taking over the conceptions of others as our own, we lose contact with the potential wisdom of our own functioning, and lose confidence in ourselves. Since these value constructs are often sharply at variance with what is going on in our own experiencing, we have in a very basic way divorced ourselves from ourselves, and this accounts for much of modern strain and insecurity. This fundamental discrepancy between the individual's concepts and what he is actually experiencing, between the intellectual structure of his values and the valuing process going on unrecognized within him—this is a part of the fundamental estrangement of modern man from himself. This is a major problem for the therapist.

Restoring Contact with Experience

Some individuals are fortunate in going beyond the picture I have just given, developing further in the direction of psychological maturity. We see this happen in psychotherapy where we endeavor to provide a climate favorable to the growth of the person. We also see it happen in life, whenever life provides a therapeutic climate for the individual. Let me concentrate on this further maturing of a value approach as I have seen it in therapy.

In the first place let me say somewhat parenthetically that the therapeutic relationship is *not* devoid of values. Quite the contrary. When it is most effective, it seems to me, it is marked by one primary value: namely, that this person, this client, has worth. He as a person is valued in his separateness and uniqueness. It is when he senses and realizes that he is prized as a person that he can slowly begin to value the different aspects of himself. Most importantly, he can begin, with much difficulty at first, to sense and to feel what is going on within him, what he is feeling, what he is experiencing, how he is reacting. He uses his experiencing as a direct referent to which he can turn in forming accurate conceptualizations and as a guide to his behavior, E. T. Gendlin has elaborated the way in which this occurs. As his experiencing becomes more and more open to him, as he is able to live more freely in the process of his feelings, then significant changes begin to occur in his approach to values. It begins to assume many of the characteristics it had in infancy.

Introjected Values in Relation to Experiencing

Perhaps I can indicate this by reviewing a few of the brief examples of introjected values which I have given, and suggesting what happens to them as the individual comes closer to what is going on within him.

The individual in therapy looks back and realizes, "But I *enjoyed* pulling my sister's hair—and that doesn't make me a bad person."

The student failing chemistry realizes, as he gets close to his own experiencing—"I don't value being a doctor, even though my parents do; I don't like chemistry; I don't like taking steps toward being a doctor; and I am not a failure for having these feelings."

The adult recognizes that sexual desires and behavior may be richly satisfying and permanently enriching in their consequences, or shallow and temporary and less than satisfying. He goes by his own experiencing, which does not always coincide with the social norms.

He considers art from a new value approach. He says, "This picture moves me deeply, means a great deal to me. It also happens to be an abstraction, but that is not the basis for my valuing it."

He recognizes freely that this communist book or person has attitudes and goals which he shares as well as ideas and values which he does not share.

He realizes that at times he experiences cooperation as meaningful and valuable to him, and that at other times he wishes to be alone and act alone.

Valuing in the Mature Person

The valuing process which seems to develop in this more mature person is in some ways very much like that in the infant, and in some ways quite different. It is fluid, flexible, based on this particular moment, and the degree to which this moment is experienced as enhancing and actualizing. Values are not held rigidly, but are continually changing. The painting which last year seemed meaningful now appears uninteresting, the way of working with individuals which was formerly experienced as good now seems inadequate, the belief which then seemed true is now experienced as only partly true, or perhaps false.

Another characteristic of the way this person values experience is that it is highly differentiated, or as the semanticists would say, extensional. As the members of my class of prospective teachers learned, general principles are not as useful as sensitively discriminating reactions. One says, "With this little boy, I just felt I should be very firm, and he seemed to welcome that, and I felt good that I had been. But I'm not that way at all with the other children most of the time." She was relying on her experiencing of the relationship with each child to guide her behavior. I have already indicated, in going through the examples, how much more differentiated are the individual's reactions to what were previously rather solid monolithic introjected values.

In another way the mature individual's approach is like that of the infant. The locus of evaluation is again established firmly within the person. It is his own experience which provides the value information or feedback. This does not mean that he is not open to all the evidence he can obtain from other sources. But it means that this is taken for what it is—outside evidence—and is not as significant as his own reactions. Thus he may be told by a friend that a new book is very disappointing. He reads two unfavorable reviews of the book. Thus his tentative hypothesis is that he will not value the book. Yet if he reads the book his valuing will be based upon the reactions it stirs in him, not on what he has been told by others.

There is also involved in this valuing process a letting oneself down into the immediacy of what one is experiencing, endeavoring to sense and to clarify all its complex meanings.

I think of a client who, toward the close of therapy, when puzzled about an issue, would put his head in his hands and say, "Now what *is* it that I'm feeling? I want to get next to it. I want to learn what it is." Then he would wait, quietly and patiently, trying to listen to himself, until he could discern the exact flavor of the feelings he was experiencing. He, like others, was trying to get close to himself.

In getting close to what is going on within himself, the process is much more complex than it is in the infant. In the mature person, it has much more scope and sweep, for there is involved in the present moment of experiencing the memory traces of all the relevant learnings from the past. This moment has not only its immediate sensory impact, but it has meaning growing out of similar experiences in the past. It has both the new and the old in it. So when I experience a painting or a person, my experiencing contains within it the learnings I have accumulated from past meetings with paintings or persons, as well as the new impact of this particular encounter. Likewise the moment of experience contains, for the mature adult, hypotheses about consequences. "I feel now that I would enjoy a third drink, but past learnings indicate that I may regret it in the morning." "It is not pleasant to express forthrightly my negative feelings to this person, but past experience indicates that in a continuing relationship it will be helpful in the long run." Past and future are both in this moment and enter into the valuing.

I find that in the person I am speaking of (and here again we see a similarity to the infant) the criterion of the valuing process is the degree to which the object of the experience actualizes the individual himself. Does it make him a richer, more complete, more fully developed person? This may sound as though it were a selfish or unsocial criterion, but it does not prove to be so, since deep and helpful relationships with others are experienced as actualizing.

Like the infant, too, the psychologically mature adult trusts and uses the wisdom of his organism, with the difference that he is able to do so knowingly. He realizes that if he can trust all of himself, his feelings and his intuitions may be wiser than his mind, that as a total person he can be more sensitive and accurate than his thoughts alone. Hence he is not afraid to say—"I feel that this experience (or this thing, or this direction) is good. Later I will probably know *why* I feel it is good." He trusts the totality of himself.

It should be evident from what I have been saying that this valuing process in the mature individual is not an easy or simple thing. The process is complex, the choices often very perplexing and difficult, and there is no guarantee that the choice which is made will in fact prove to be self-actualizing. But because whatever evidence exists is available to the individual, and because he is open to his experiencing, errors are correctable. If a chosen course of action is not self-enhancing this will be sensed and he can make an adjustment or revision. He thrives on a maximum feedback interchange, and thus, like the gyroscopic compass on a ship, can continually correct his course toward becoming more of himself.

Some Propositions Regarding the Valuing Process

Let me sharpen the meaning of what I have been saying by stating three propositions which contain the essential elements of this viewpoint. While it may not be possible to devise empirical tests of each proposition in its entirety, yet each is to some degree capable of being tested through the methods of science. I would also state that though the following propositions are stated firmly in order to give them clarity, I am actually advancing them as decidedly tentative hypotheses.

1. *There is an organismic base for an organized valuing process within the human individual.*

It is hypothesized that this base is something the human being shares with the rest of the animate world. It is part of the functioning life process of any healthy organism. It is the capacity for receiving feedback information which enables the organism continually to adjust its behavior and reactions so as to achieve the maximum possible self-enhancement.

2. *This valuing process in the human being is effective in achieving self-enhancement to the degree that the individual is open to the experiencing which is going on within himself.*

I have tried to give two examples of individuals who are close to their own experiencing: the tiny infant who has not yet learned to deny in his awareness the processes going on within; and the psychologically mature person who has relearned the advantages of this open state.

3. *One way of assisting the individual to move toward*

openness to experience is through a relationship in which he is prized as a separate person, in which the experiencing going on within him is emphatically understood and valued, and in which he is given the freedom to experience his own feelings and those of others without being threatened in doing so.

This proposition obviously grows out of therapeutic experience. It is a brief statement of the essential qualities in the therapeutic relationship. There are already some empirical studies, of which the one by Barrett-Lennard is a good example, which gives support to such a statement.

Propositions Regarding the Outcomes of the Valuing Process

I come now to the nub of any theory of values or valuing. What are its consequences? I should like to move into this new ground by stating bluntly two propositions as to the qualities of behavior which emerge from this valuing process. I shall then give some of the evidence from my own experience as a therapist in support of these propositions.

4. *In persons who are moving toward greater openness to their experiencing, there is an organismic commonality of value directions.*

5. *These common value directions are of such kinds as to enhance the development of the individual himself, of others in his community, and to make for the survival and evolution of his species.*

It has been a striking fact of my experience that in therapy, where individuals are valued, where there is greater freedom to feel and to be, certain value directions seem to emerge. These are not chaotic directions but instead have a surprising commonality. This commonality is not dependent on the personality of the therapist, for I have seen these trends emerge in the clients of therapists sharply different in personality. This commonality does not seem to be due to the influences of any one culture, for I have found evidence of these directions in cultures as divergent as those of the United States, Holland, France, and Japan. I like to think that this commonality of value directions is due to the fact that we all belong to the same species—that just as a human infant tends, individually, to select a diet similar to that selected by other human infants, so a client in therapy tends, individually, to choose value directions similar to those chosen by other clients. As a species there may be certain elements of experience which tend to make for inner development and which would be chosen by all individuals if they were genuinely free to choose.

Let me indicate a few of these value directions as I see them in my clients as they move in the direction of personal growth and maturity.

They tend to move away from facades. Pretense, defensiveness, putting up a front, tend to be negatively valued.

They tend to move away from "oughts." The compelling feeling of "I ought to do or be thus and so" is negatively valued. The client moves away from being what he "ought to be," no matter who has set that imperative.

They tend to move away from meeting the expectations of others. Pleasing others, as a goal in itself, is negatively valued.

Being real is positively valued. The client tends to move toward being himself, being his real feelings, being what he is. This seems to be a very deep preference.

Self-direction is positively valued. The client discovers an increasing pride and confidence in making his own choices, guiding his own life.

One's self, one's own feelings come to be positively valued. From a point where he looks upon himself with contempt and despair, the client comes to value himself and his reactions as being of worth.

Being a process is positively valued. From desiring some fixed goal, clients come to prefer the excitement of being a process of potentialities being born.

Perhaps more than all else, the client comes to value an openness to all of his inner and outer experience. To be open to and sensitive to his own *inner* reactions and feelings, the reactions and feelings of others, and the realities of the objective world—this is a direction which he clearly prefers. This openness becomes the client's most valued resource.

Sensitivity to others and acceptance of others is positively valued. The client comes to appreciate others for what they are, just as he has come to appreciate himself for what he is.

Finally, deep relationships are positively valued. To achieve a close, intimate, real, fully communicative relationship with another person seems to meet a deep need in every individual, and is very highly valued.

These then are some of the preferred directions which I have observed in individuals moving toward personality maturity. Though I am sure that the list I have given is inadequate and perhaps to some degree inaccurate, it holds for me exciting possibilities. Let me try to explain why.

I find it significant that when individuals are prized as persons, the values they select do not run the full gamut of possibilities. I do not find, in such a climate of freedom, that one person comes to value fraud and murder and thievery, while another values a life of self-sacrifice, and another values only money. Instead there seems to be a deep and underlying thread of commonality. I dare to believe that when the human being is inwardly free to choose whatever he deeply values, he tends to value those objects, experiences and goals which make for his own survival, growth, and development, and for the survival and development of others. I hypothesize that it is characteristic of the human organism to prefer such actualizing and socialized goals when he is exposed to a growth-promoting climate.

A corollary of what I have been saying is that in *any* culture, given a climate of respect and freedom in which he is valued as a person, the mature individual would tend to choose and prefer these same value directions. This is a highly significant hypothesis which could be tested. It means that though the individual of whom I am speaking would not have a consistent or even a stable system of conceived values, the valuing process within him would lead to emerging value directions which would be constant across cultures and across time.

Another implication I see is that individuals who exhibit the fluid valuing process I have tried to describe, whose value directions are generally those I have listed, would be highly effective in the ongoing process of human evolution. If the human species is to survive at all on this globe, the human being must become more readily adaptive to new problems and situations, must be able to select that which is valuable for development and survival out of new and complex situations, must be accurate in his appreciation of reality if he is to make such selections. The psychologically mature person as I have described him has, I believe, the qualities which would cause him to value those experiences which would make for the survival and enhancement of the human race. He would be a worthy participant and guide in the process of human evolution.

Finally, it appears that we have returned to the issue of universality of values, but by a different route. Instead of universal values "out there," or a universal value system imposed by some group—philosophers, rulers, or priests—we have the possibility of universal human value directions emerging from the experiencing of the human organism. Evidence from therapy indicates that both personal and social values emerge as natural, and experienced, when the individual is close to his own organismic valuing process. The suggestion is that though modern man no longer trusts religion or science or philosophy nor any system of beliefs to *give* him his values, he may find an organismic valuing base within himself which, if he can learn again to be in touch with it, will prove to be an organized, adaptive and social approach to the perplexing value issues which face all of us.

Summary

I have tried to present some observations, growing out of experience in psychotherapy, which are relevant to man's search for some satisfying basis for his approach to values.

I have described the human infant as he enters directly into an evaluating transaction with his world, appreciating or rejecting his experiences as they have meaning for his own actualization, utilizing all the wisdom of his tiny but complex organism.

I have said that we seem to lose this capacity for direct evaluation, and come to behave in those ways and to act in terms of those values which will bring us social approval, affection, esteem. To buy love we relinquish the valuing process. Because the center of our lives now lies in others, we are fearful and insecure, and must cling rigidly to the values we have introjected.

But if life or therapy gives us favorable conditions for continuing our psychological growth, we move on in something of a spiral, developing an approach to values which partakes of the infant's directness and fluidity but goes far beyond him in its richness. In our transactions with experience we are again the locus or source of valuing, we prefer those experiences which in the long run are enhancing, we utilize all the richness of our cognitive learning and functioning, but at the same time we trust the wisdom of our organism.

I have pointed out that these observations lead to certain basic statements. Man has within him an organismic basis for valuing. To the extent that he can be freely in touch with this valuing process in himself, he will behave in ways which are self-enhancing. We even know some of the conditions which enable him to be in touch with his own experiencing process.

In therapy, such openness to experience leads to emerging value directions which appear to be common across individuals and perhaps even across cultures. Stated in older terms, individuals who are thus in touch with their experiencing come to value such directions as sincerity, independence, self-direction, self-knowledge, social responsivity, social responsibility, and loving interpersonal relationships.

I have concluded that a new kind of emergent universality of value directions becomes possible when individuals move in the direction of psychological maturity, or more accurately, move in the direction of becoming open to their experiencing. Such a value base appears to make for the enhancement of self and others, and to promote a positive evolutionary process.

I

From My Life

Barry Stevens

> "He (the infant) would laugh at our concern over values, if he could understand it. How could anyone fail to know what he liked or disliked, what was good for him and what was not?"

When I read that, I remembered my friends saying to me so often when I was young, "You're lucky. You always know what you want." I thought a person must be crazy not to.

When I was forty years old, I was baffled and confused because I couldn't seem to know what I wanted. In my own terms, I had gone crazy.

In trying to find my way out of this, I went two ways at once: a search inside myself for what had gone wrong, and a search outside myself for something to believe that would set me right. The outside search was a flop. I never did find anything that I could entirely go along with. The inside search was rewarding, and it was there that I found that I didn't need to believe anything at all. Everything that I needed was right inside me. The outside was useful only when it helped me to get in touch with what was inside. But when I did get in touch with my inner valuing again, it was terribly hard to trust it, because in important ways it went against what everyone says. The more I use it, the more I trust it, and when I am really close to other people, I find that their built-in pathfinders (my term for what Carl Rogers calls "organismic valuing") agree with mine. The difference between the outside and the inside view goes like this:

When my son was at college, he got picked up for driving

a Model A roadster with an excess number of passengers, some of them on the running board, and was fined $27.00. This hurt. He had worked a good deal since the age of nine. At college, he had a tuition scholarship but was otherwise supporting himself by having several jobs and he was helping me too, as I was sick in bed for several years. To him, $27.00 was more than three days' labor. It hurt to pay the fine, but he did not resent this. He knew what the law was, and knew that he had broken it. He accepted his own responsibility for what happened.

But at the police station he was told that he was irresponsible. This really seared him. He was made to feel *bad,* and that is not good. He also felt wronged and misjudged, and this made him very resentful. At the same time, he was confused, which is probably worse than anything else. Several years later, when he was in graduate school in another state, two policemen came to our door asking for money for the Fourth of July fireworks display. We really loved those fireworks, and he freely gave them five dollars although we didn't have much money then, either. After they had gone he said, "I still hate cops. When I see them, I feel it."

As I see it, he was not irresponsible. He was driving the other boys only two blocks from a dorm to the athletic field, in an area where there was very little and no fast traffic. He was alert to the fact of the young men on the running board and he also knew their own alertness and ability to look after themselves. He had made *himself* responsible. To me, it is not responsible to drive sixty miles an hour in accord with the speed limit when stretches of the road have become unsafe for driving at that speed, or weather conditions make it hazardous. A person who does that goes exclusively by the rule, instead of including his own noticing, his own awareness, and when there is a wreck he is sure he has "done nothing wrong." The bad road was what did it, or the weather. It seems to me that I am responsible when I am responseable to everything around me, and that the opposite is the Eichmanns who have "done nothing wrong" because they did what they were told to do.

When I had been in the hospital for a month, five days after leaving it I had to go to see the doctor whose office was in a private clinic. While I was waiting to see him, I realized that I was slipping from the chair, and the only way that I could keep myself in it was to hold tightly to the arms. I wasn't sure how long I could hold on, and noticed that I was getting woozy in the head. I got myself up and to the desk, where I had to lean over the chest-high counter and hook my fingers on the opposite edge to keep myself from slipping to the floor. I told one of the nurses that I needed to lie down. She asked me, "Who is your doctor?" "Do you have an appointment?" "What is your clinic number?" What did any of that have to do with a sick person *who needed to lie down?* That must have been obvious quite apart from my telling her so. She was neither a cold person nor a wicked one, nor in most respects was she stupid. She had made herself a "responsible" person who abides by the rules, and her "responsibility" was to her *job* as it was defined by the administration, not to the immediate need of another human being. Like Eichmann.

My predicament in the clinic is so prevalent in our society that I think there can't be anyone who hasn't run into something similar—perhaps hundreds of times—plus all the others that one hears about, like the child who was brought to a police station because he had been bitten by a rattlesnake, and he was left sitting there while the cops tried to find out where his home was, so they'd know which hospital to send him to. Everybody knows about things like this, but nobody *does* anything about it. It can really make a person scared.

Sometimes the same kind of thing leads only to an absurdity. During the war, in southern California, my daughter wanted to take a course in aircraft production illustration. The man in charge said he couldn't accept her because she wouldn't be eighteen by the time she was through with the course. That was the rule. I went to see him, and he was really a swell person, but he wouldn't budge even when I said, "Look. Here's a girl who is very good at drawing and she loves planes. She has come from a war zone and wants to do something that doesn't seem futile to her. She's just what you need, so why not admit her?" He said, "Oh, I couldn't do that! I'd have to go over several dead bodies." I thought of the bodies dumped into trenches in Honolulu, and the bodies left rotting in ships in Pearl Harbor because there wasn't time to do anything about them. His remark in this context was too much for me. I said, "In a war, what's a few more dead bodies?" That was too much for him, too. He got her in.

When my husband was in charge of the pediatric service

at a hospital in New York in the twenties, there was an infant whom none of the doctors could find anything wrong with, but all of them agreed that the baby was dying. My husband spoke privately to a young nurse who loved babies. He swore her to secrecy before telling her what he wanted her to do. The awful secret was, "Take care of this baby as though it were your own. Just *love* it." At that time, "love" was nonsense even to psychologists; to doctors and nurses, it still seems to be what you mustn't have for a patient. The baby took hold. All the doctors agreed on that. But if my husband had told the other doctors how this happened, he would have ceased to be a medical man (trustworthy) and would have become a mystic (unreliable). Even if some of his colleagues might have agreed with him, they wouldn't have dared to speak in support of him because then they would have lost caste too. Love was not "scientific" because it could not be measured. So let the baby die?

Two quite well-known scientists have told me, separately, of things they had observed about life, their own knowing, and when they were leaving said, in identical words, "Don't tell anyone I said that!"

A psychologist said one thing at school and another within his own home. When asked about the discrepancy he said, "That was my professional opinion (at school). This is my personal opinion." If *schizophrenic* means "split mind," then who isn't? No wonder that when William Menninger was asked how many of us suffer from emotional illness he answered, "One out of one of us."

I have never been a doctor, a scientist, or a professor, but the struggle between acting on my own authority and going along with outside authority when this is not in accord with me continues, even though I now know clearly which authority it is that I value. It hurts deeply to be told that I am irresponsible—like a knife thrust into my chest and given a twist. So I know somewhat how it must feel to professional people, and why they don't speak out more than they do. When I do and say what everyone says and does, then no one calls me irresponsible. But at that time, I am.

I am inconsistent, not congruent, (sometimes called a hypocrite) if I complain about bribery and deceit in politics, government, business, the police, when I myself do and say, at the expense of my own integrity, what I will be rewarded for in smiles, friendship, acceptance, position, a nice house and all the other things which are supposed to be our good and proper goals. The wickedness is not in what I have accepted, but in what I have given up, which is myself, my own authority based on my own knowing. This process begins so early in our lives even under relatively good conditions that I can't blame anyone else or me for becoming confused, but no matter who got me into what I got into, *I* am the only person who can get me out of it. Others can certainly help—and they have—by letting me think what I think, choose what I choose, and feel what I feel. But still, *I* have to be willing to let this surge into me and become the basis for my actions. This can be ridiculously difficult and frightening. It may be about something that doesn't seem in the least important when looked at from the outside, but the inside scene is altogether different.

My husband, as a part of the age he lived in and his profession, was contemptuous of "mysticism." This was the same man who cured a baby by assigning it to a loving nurse. His feeling about swamis and ochre robes—of which he had no direct experience whatever—was so strong that when Aldous Huxley, whom he had admired, joined the Vedantists my husband said bitterly, "Git along, little yogi." I had got infected by his shudders. When I was on the mainland with the children for a year and started out in search of my own values, I went to the Vedanta temple in Hollywood to find out for myself what I thought of swamis and ochre robes. That is true, and yet the way that is stated is misleading. It expresses a clarity which was not present at the time. A more accurate way of saying it is, "I didn't know what I was doing but I knew that I had to do it"—the wisdom of the organism of which Carl Rogers speaks, making its own corrections. I sat through the service with ants running up my back. I felt that I must have gone out of my mind to be there, because I didn't know anyone who wouldn't disapprove of me. Something made me stay, not run away. Afterward, although I had been a devotee of the non-handshaking cult for many years, I went to the swami and shook hands with him. I didn't know why: it was just something that I had to do. As I looked at him, suddenly I felt very shaky and my voice cracked as I said, with deep and genuine feeling, "Thank you!" I felt a fool for my shakiness and my emotion, but it made no difference to the swami: there was no change in him. His acceptance of me was the same before, during, and after. I didn't know what I was thanking him for until I realized that I *liked* the guy. He was *real*. His being real, not phoney, had helped me to break through what had blocked me, which was such a battle taking place in me that it felt like exorcising the devil, like breaking out of a strait-jacket. But somehow I got out. I'm sure there are phoney swamis just as there are phoney everything else—ministers, lawyers, doctors, teachers, scientists—but *this* one *wasn't*, and I could never have that awful block against a whole group—swamis—again. That was my only face-to-face encounter with a swami and almost my total experience of him. I had lunch once with a swami recently arrived from India—a very sincere young man who was also very nervous. Another swami, I listened to for an hour at a lecture and kept looking at my watch. I haven't since been able to think of "swamis" *as* anything, but only of individual swamis, and this I like because it's real. I have never yet known anyone who fit a category or who was *only* the category in which he was placed. Some were worse, some better, but the category itself was misleading.

At the same time, I was left *open* to "mysticism." I didn't accept it, but neither did I block it out. I was free to explore it or not, but I knew that I couldn't say anything *about* it until I *had* explored it—until I could speak from my own experience and know what *I* was talking about. This seems to be part of the built-in pathfinder, that it finds its *own* way regardless of what anyone else says or thinks. It acts on the information that it has, but tentatively—open to change as further information comes in. Irrational as it seems to my rational mind, it is—in terms of my own life—*more* scientific. It explores, discovers, tests, is forever open to re-evaluation and perpetual learning. It doesn't get fuddled or irritated by mistakes: it is interested in what happens, learns, moves on. It is not "coldly scientific" any more than Nobel prize-winning scientists like Linus Pauling and Albert Szent-Gyorgyi are "cold"—they're warm, human, enthusiastic, don't take themselves too seriously and are very much alive.

One needs only to watch a healthy infant or small child, forever testing and exploring and enjoying this, to know what I have rediscovered in myself.

When I was small, one of the things that puzzled me was

that when I saw something that I wanted to try, and did it, sometimes the grownups said that I was bright, sometimes that I was silly. A little later, with people outside the family who did not love me as my family did, it was sometimes I was "bright" and sometimes I was "stupid." I couldn't understand at first what made the difference. As I went into doing things, they looked the same to me. Gradually I learned that "bright" or "silly" depended not on how it looked to me when I went into it, but on *how it came out.* That was puzzling to me, because how it came out was something that I never knew until after I had done it. I did things *to see* what would happen. So how could I be "bright" when it came out one way, and "silly" when it came out another? *I* was the same both ways, it seemed to me.

Later on I learned, especially in school, to value "success" and to hide "failure" so that I wouldn't be scolded or ridiculed. That wasn't the way that I started out, when *both* were interesting, and failure was sometimes more stimulating than success because it raised more questions. When I turned my mind to concealing failure—being clever about it—I didn't notice the questions any more.

When Carl Rogers writes of values "introjected from other individuals" but "regarded by him as his own," he refers to a major conflict, which sometimes becomes a bloody battleground. I made the usual mistake of thinking that everything that went on in my head was me. Even when I was 56 and had got a good deal of that cleared up, I got into an aching mess for a week about something having to do with money. The back of my neck was clutched, and my head ached unbearably because there were two opposing views fighting in my head and I couldn't settle for either one. I was so tired of fighting that I didn't care *which* side won, if only the blasted battle would stop. But no matter which side I decided to choose, I felt guilty. Then I had a dream which, speaking its own language, told me that the conflicting views of money were neither of them mine. It was really a battle between my father, who never forgot a debt whether it was his own or another's, and my uncle who never remembered one unless he was reminded and then it seemed to him unimportant. My *own* view of money was neither. What *I* thought about money had to do with this particular instance which had some unusual factors in it. It was fantastic to me that I had worn myself out for a week in a battle that was really between two men with whom I had spent very little time for twenty years, and both of whom were dead.

When Carl Rogers speaks about "prizing" a person, and conveying the feeling "this person has worth," I become uneasy that this may be misconstrued to mean a kind of praise, of placing a *high* (selective, comparative, superior) value on a person. To me, it means not that, but something which is more difficult to describe, something that is not praise, not blame, and at the same time not neutral, flat, or middle-of-the-road. To me it is closer to the equality that I learned from Herbert Talehaftewa, a Hopi who at home on the Reservation was a kind of circuit judge. He was working as a carpenter on a construction job where I was office manager. Cab, the owner and boss, was a Boston snob who looked down on *everyone,* belittled them to the point where most people who were subjected to it went to pieces and had to pull themselves together again. One day I saw this man look at and speak to the Hopi in this way. Cab was a small man, and the Hopi was quite tall and broad, but Cab managed to look down on the Hopi. I saw the Hopi look at Cab so *equally* that he drew Cab down to his own level—precisely, and not one bit lower—so that they seemed to be two people eye-to-eye. I was so impressed by this that I looked up to the Hopi as though he were some sort of god. The Hopi turned to me with that same strong *equalness* in his gaze, and I felt myself being drawn *up* until we were on the same plane. Through him I knew that all men are equal if only we will regard them so.

This *equalness* is what "prizing" and "this person has worth" say to me—not exceptional, although at the same time unique, but *equal* with me myself who also is not exceptional and still has worth and is unique. "You are meaningful to me as one person to another." "You are as interesting to me as I am to myself." Differences in physique, sex, dress, speech, age, education, background—all of these disappear in the sense that although they are present, they are unimportant. We are in direct communication with each other—person to person.

At the time of the incident with the Hopi, my office manager work was partly in abeyance because I was cooking three times a day for a dozen Hopi men who worked on construction. The Tewa Indian cook had burned her hand severely and had to stop using it for ten days. I hoped that my cooking was pleasing to the Hopi men, but I didn't know. One day Herbert Talehaftewa, the circuit judge at home, said to me evenly, "The men say you are doing the best you can." I was hurt. It seemed to me that they must think that my cooking was not very good. But then I realized that what the men said was simple truth, and that their recognition of that was more beautiful to me than praise. They knew me *innerly*. And isn't that the way that all of us wish to be known, no matter how many blocks and barriers we may put up against it?

I came to know them innerly too, person to person, and fifteen years later, with only scattered messages in between, when I hear of something that has gone well or ill in their personal lives, I feel this deeply within myself, knowing truly what it means to them, to each man in terms of *himself*, and at the same time in terms of all of us—the whole human race. I am closer now to those Hopi men whom I haven't seen for fifteen years than I am to many of the people now around me who have categorized me, put me in a pigeon-hole, who do not know me *innerly* at all. I wouldn't hesitate to tell any of those Hopi men my troubles, of any kind, because they would simply accept them, not try to advise me, and their acceptance would be in a sharing way, without regard to differences.

A year ago at New Year's I was talking with a Navajo man who had said "Happy New Year" to me. I asked him if he thought the New Year would be good for him or if it would be like my own life, "I think there is no end to being down—and then things change. They go along well, and I think, 'now at last, everything is all right'—and then it isn't." He nodded, and said simply, "Just like us." There was complete acceptance that life *was* the same for both of us, even though I had never been hungry as he had, had never been in jail as he had, and had never been forced to submit to the imposition of another culture as he had. That this is true, I know through listening to a woman talk for seventeen hours, in three days, about her life. Her life had been as "different" from mine as mine was from the Navajo's. We had very little and most of what we had was second-hand. This woman had been born to Old Masters, a yacht with a crew of twenty-eight men, and everything that went with that. The more I listened, the more I knew that there had been no difference between the lives of the Princess and the Pauper. The *innerness* was the same.

The "commonality" of which Carl Rogers speaks means to me what I have found in all of us, without exception. Sometimes it is clearly the mainstream of a person's life even though smaller streams confuse it. Other times, it has showed only briefly in a person when through stress or ease the barriers went down. This used to be a puzzle to me: which was the person? Other people said that what a person was most of the time was what he *was,* and this confused me, because it seemed reasonable and yet—what the person was at times in privacy seemed so much more vibrant, living, *real.* Sometimes his pouring out of it was almost more than I could bear, the agony of man's knowing his inability to be what he knew himself to be. Sometimes it was gentle and tender, this knowing, and so sad.

Which *is* the person? A woman told me that one day she was at a football game waving a pennant and cheering, when suddenly the hand holding the pennant fell to her lap, and with all the excited mob around her she thought, "Why am I doing this? I haven't *enjoyed* it for fifteen years." Which was real—the fifteen years or the moment?

The knowing of the realness in another person—his desperate desire to be non-hurtful, loving, responsive, constructive (or creative), at one with others—seems to me to explain why one person sometimes stays with another against all reason, and not necessarily with good judgment. A very simple "ignorant" woman, who had every reason to hate her husband, told me with tears, appealing to me with all her being to know this, "He wants to be good." Mothers often know this in their children and defend *them*—not their behavior—when they are destructive.

To yearn for something is to value it. I have known this yearning "to be good" in all of us. No one is left out. I have known it in a British peer. I have known it in a man who hijacked a truck, got an Indian drunk and threw him off the moving truck, hit a burro and killed it, bashed in the radiator, and then drove the truck until the motor burned out and he landed in jail, while four small children were at home alone because their mother was in the hospital having a fifth one. I got him out of jail so that he could take care of his family, and he got me fired from my job. I also knew it in my husband, who killed himself in 1945. Before his suicide, I lived with knowing that he might kill our son and me, both of whom he loved, and that he might seriously injure other people, which he did not wish to do at all. It seemed to me that suicide might have seemed to him the only way to put an end to his hurting people, to doing what he didn't want to do.

I know that there is this commonality. I have wanted it all my life and I was always told that it was impossible. But in December 1941, I knew the fact of living with it, and how good it is when everyone responds in terms of his own humanness, uncontaminated by all the phoney values with which we live. (The only exceptions were two very neurotic men and a few feeble-minded oldfolks who didn't know what was going on.) In Hawaii during the week following the Japanese attack on Pearl Harbor, the military expected that the success of the attack would be followed by invasion: they believed that when the Japanese planes reported to their carriers, then other ships would move in with troops. We knew that we were knocked flat, that we had nothing to fight with, and the rumor was that the military had said that a mere 2,000 men could take the island. Where I was living was within the twenty-mile area where the invasion was expected to take place. We didn't expect to live through it.

All appointments of every kind had been cancelled. All routines were so suddenly wiped out that when on that Monday morning I went out the door and saw bottles of milk standing on the steps, I said, "The milk!" with astonishment that anything could be as it had been before.

Everywhere I went, the people—Japanese, Chinese, Hawaiian, Portuguese, *haole* (white)—seemed to be walking around with halos over their heads that they didn't know were there. "Wouldn't you like this?" (as though, "I have too much") "Isn't there something that I can do for you?" The gentle asking, truly meant, but at the same time without insistence.

It is difficult for me to go on writing this. Tears blur my vision, no matter how many times I wipe them away. I sob, and that shakes me so that I can't use the typewriter very well. I feel, "Oh, what's the use? How can I possibly convey it?"

Some people went out with children, harvesting macadamia

nuts. It didn't matter to anyone whose land they grew on, or who ate them.

In a little Japanese grocery store where Mr. Yoshimoto and I had discussed everything from the difficulty of getting good coffee in Japan to international affairs, Mr. Yoshimoto looked up as I came in and said, "I think now you and Mr. Lindbergh no can talk." We laughed together. The other Japanese in the store laughed companionably with us, except one man who had come in just behind me. He was twirling around his finger the shoelaces that were in a box on a counter. Mr. Yoshimoto asked him, with the intuition that was quick in all of us then, "Your son?"

"Me no know," said the other man. "He take suitcase. Pearl Harbor go." He went on twirling the shoelaces around his finger.

All of us were silent—not "showing respect" or "trying to feel for someone else." I had never seen this man before. I knew only that he meant that after the first attack, when all civilian workers were asked on the radio to come to Pearl Harbor at once, his son had gone. We had heard that sentries at the gate had not heard of the appeal, and when men, many of them Japanese, came driving up to the gate fast, the sentries had fired. There was also the second attack.

But what this man *felt* about his son was felt and known by all of us. It can't be described. But how would you feel? That was our knowing as we stood unmoving in the utter silence. Even my six-year-old son was still.

The swift transition that took place then, from our laughter to our silence, was happening all the time as we responded to each other everywhere. So much was funny, so much was sad. A glint of laughter in an eye was gone along with, merrily, or washed away at the hint of something in the eye or body posture of someone else. A hand would reach out and touch someone, not knowing why, and something was shared.

A mother would start to scold a child, then the thought, "What does it matter?"

A mother would be gentle with a child with a special painful tenderness, then notice fear in the child's eyes, and with a swift transition be firm or laughing, and the child's eyes would clear.

All that week, I didn't know anyone who pretended he

wasn't afraid, except to children. There were soldiers with guns posted here and there, standing behind a few sandbags that came not much above their ankles—never to their knees. That seemed to symbolize the degree of our protection. We felt that we had none. When we heard a plane, it was almost certain to be Japanese. It seemed to me that I would be so happy in an igloo in Alaska, if only I were away from war.

And yet, at the same time, there was all this living for which I had always longed—the beautiful feeling of people everywhere, the sharing, the knowing, the responsiveness, the free willingness and nonpossessiveness of things or other people.

I was not on the Honolulu side of the island all that week, so I don't know how it was there, but an officer in the army engineer corps told me that during that week merchants everywhere were opening their stores to the military to take whatever they needed, with a feeling of real friendliness and no accounting—"And then, on Monday a week later, when the immediate scare was over . . ." His lips were tight and he said with sadness for what had been lost, and with a touch of bitterness, "they began sending in the bills—at wartime prices." On our side of the island then, almost everyone was saying they hadn't been scared—the same people who, the week before, had told me that they were so terribly afraid.

The happiness happened when we were scared, but it wasn't *only* or even necessarily the being scared. People can do the most frightful things when they are scared. But we expected to be wiped out at any moment, and there was only now to live in. That's what we did.

It seems to me that is what Carl Rogers does when he is with his clients, and that it's about time we all did. We pay so much attention to *where* we live, so little to *when.*

II

> Life has proceeded under the guidance of religion, science, morality and economics; it has even proceeded under the capricious direction of art or pleasure; the one expedient that has never been essayed is that of living intentionally under the guidance of life. Fortunately, mankind has always more or less lived in this way, but such living has been unintentional; as soon as men saw what they were doing they repented, and experienced a mysterious remorse.
>
> Jose Ortega y Gasset*

To be or not to be. In my beginning, that was no question. It has become a silly question to me now. But in between, I was as confused as Look Before You Leap and He Who Hesitates Is Lost.

In the beginning, so clear. Even when I had to yield to superior force and do what is not in accord with life, which includes *my* life, this was done clearly for the time being only, until I should become a grownup myself and my life would be my own. When that happened,

"A rolling stone gathers no moss," they said.

"Who wants moss? And look at a snowball." There is so much for me to know and learn.

"You *must* have a career," they said.

"I don't *want* a career," I said. "I want a careen," I said, with a sense of the world before me, and knowing that for me a straight line through it would miss too much.

---

* From *The Modern Theme,* by Jose Ortega y Gasset. Harper Torchbook, New York, 1961, p. 61.

When I was young, the remorse was the other way around from that which Ortega saw—remorse for the times when I wasted life doing what others wanted me to, like chattering with chattering people about other people's chatter. When I did that, I woke up with a bad taste in my mouth and thinking sadly, "That ain't none of I."

I liked better to sit sometimes and listen to a toothless old woman, or some drunk, talk about life as it was to them, because that was what I wanted to know, and most people wouldn't tell me. But when I was seen doing this, it was "bad." Sometimes it seemed to me that everything that was "good" to me was "bad" in the eyes and minds of others. I did something feeling so pure, and then was told that it was disgusting. And eventually I got switched around too—the label "bad" now stuck on what was good.

Then, when I had got reversed, I knew remorse when I had been forgetful and did what felt good. I thought that what had felt good was bad, and that I shouldn't have done it. This was indeed a mysterious remorse. It is what people call "becoming mature." That was the way I thought of it when Other I took over and wiped out the I which is *me.*

A woman on a train, crossing Arizona at two o'clock in the morning, said to me with sighs, "It's awful when you find out that everyone is insane and you are too. By that time, the younger generation is already into it, and your capillaries have broken down and you can't do anything about it." "You know, Mrs. Stevens," another woman said, referring to herself with anger, "You can sophisticate your mind so, that you believe lies!"

But young people, already switched around to lies, or "into it," attribute the distress of old folks like us to other problems: we're warped because we're widows, or because our husbands have proved disappointing (if one explanation doesn't fit, another must be found) but *their* lives are never going to be like that. (Generation after generation affirms this silently, or whispers it, or shouts it, with such assurance.)

But now, there is the beginning of a change. Some people aren't waiting until they're forty. A woman in her early thirties writes, "Well, adulthood seems to consist of shoveling out the manure that accumulated in childhood. It certainly takes a strong constitution to keep at it." And another, ten years younger, says "I'm beginning to see where I foul up other people. I'm not so good, yet, at seeing where other people foul up me." My son, still in his twenties, listened to a child psychology film for awhile and then reversed it. As he did so, "The grownups' voices came out backwards, but the child's cry, somehow, seemed to come out *frontwards*, and I had the thought, 'Suppose the child starts out talking frontwards and we spend twenty years teaching him to talk backwards?' "

"I hate my mother."

"Oh no, you don't. You *love* your mother."

In the first grade, questions start sparkling in my mind. I ask them and am told, "You are disrupting the class." At first, I don't understand the word "disrupting" but by its being said over and over I get the hang of it. I love the questions which disrupt the class, so then I think of me, in second grade, "I *like* to disrupt the class." And then I do throw in questions *to* disrupt the class, when in the beginning all I wanted was to explore the questions.

My life goes well. I do not do much thinking about it, but when something is wanted or needed I find my way to it. "You're a wonderful manager!" Over and over and over, year by year, and then I thought that I *must manage*, to be a good manager, and wondered why it became so much more difficult to do what earlier had been done with ease.

"You *must* think first."

"But it came out fine."

"It might *not* have," and all the terrible things that *might* have happened are explained to me, and I am told that I must look into all the possibilities and not take the chances that I do. And then, when I had changed my life to something to be controlled by me (or *managed*), instead of something that I went along with, it was only when I was forgetful that I lived under the guidance of life, and this was followed by remorse.

More and more it seemed to me that my life was swallowed by always doing first what I MUST do, with what I *wanted* to do relegated to whatever time and energy were left, and the time and energy were constantly diminished, so that I wore myself out trying to get through what had to be done so that there would be a *little* time left for what I wanted to do.

Somehow, first and second places or importances had got switched around—all over. The woman on the train in Arizona said "Now, a child can't even run and feel the wind. He's got to read about it in a book first, and then he never knows."

"Yes, darling, when I've done the dishes. . . ."

Always later, later, later ("when I retire") for life and living, for what sings in me in the beginning and later fizzles to a whimper. Pie-in-the-sky laughed out of the hereafter and accepted here on earth as making sense.

"Can't we ever live *now?*"

By the time the Four Freedoms were announced, I couldn't get excited about them because I had those freedoms, and no matter how free anyone else thought I should feel because I had them. I *didn't* feel free. Freedom was something that I remembered having had. I didn't know how I had lost it, and had no idea how to get it back. When I spoke of this to people, seeking help, they said, "Of course. . . ." "That's the way life is." "It's that way for everyone." "You have to accept it." Or, "The menopause is always a difficult time." (The menopause came twelve years later and gave me no trouble.)

I couldn't believe it. What point would there be to living if that were all there was to life? An atom bomb would be a happy ending.

Starved for what I wished of life, for what I needed in the world, there seemed to be only one thing that I could do: I must put it there myself. I must take freedom so that I could give freedom. In taking it, I put it in the world as much as when I give. I was very happy when I discovered that. It was such a switch from "you can't win for losing" which had seemed to be a rule. In taking freedom I discovered that life does have certain principles and that when I live in accord with them there are only limits which do not interfere with me. They are not *felt* as limits, any more than I find it limiting to have only one nose. My skin limits me too. I find it comfortable to be inside it.

When I am within this freedom, it is silly to think of "learning" ethics. Ethics course through me with my blood, without my thinking. What we call ethics simply *is*—a part of mankind, existing as a free response to life and to all living. How can I wish to hurt you, when my feeling is that you are me and I am you? How can I feel grateful that you give to me when I am you? There is only happiness in the giving that is receiving too—happiness without responsibility or obligation. It is the happiness of responding to myself, in you, in me, without a dfference—the happening of myself, unhindered by the small mind's notions. I can no more *make* myself happen than I can beat my heart: I can only remove what prevents myself from happening. When I do this, I am learning to be free.

*Learning to Be Free**

Carl R. Rogers

I would like to describe for you a pattern of experience which I have observed, and in which I have participated. It is an experience which I have seen repeated with many variations, in many individuals, with many differing outward expressions, but with a seemingly common core. It is an experience on which I have placed various labels as I have tried to think about it—becoming a person, freedom to be, courage to be, learning to be free—yet the experience is something broader than, and deeper than, any of its labels. It is quite possible that the words I use in regard to it may miscommunicate. The speculations and ideas I present, based on this experience, may be erroneous, or partly erroneous. But the experience itself *exists*. It is a deeply compelling phenomenon for anyone who has observed it, or who has lived it.

The Experience of Learning to be Free

The experience to which I am referring is a central process or central aspect of psychotherapy. It is the experience of becoming a more autonomous, more spontaneous, more confident person. It is the experience of freedom to be one's self.

In the relationship with an effective therapist—and I shall have more to say about the qualities of this relationship—the client moves gradually toward a new type of realization, a dawning recognition that in some sense he chooses himself. This is not usually any sudden burst of insight—it is a grop-

* From *Conflict and Creativity* edited by Farber & Wilson.

ing, ambivalent, confused and uncertain movement into a new territory. The client begins to realize, "I am not compelled to be simply the creation of others, molded by their expectancies, shaped by their demands. I am compelled to be a victim of unknown forces in myself. I am less and less a creature of influences in myself which operate beyond my ken in the realms of the unconscious. I am increasingly the architect of self. I am free to will and choose. I can, through accepting my individuality, my 'isness,' become more of my uniqueness, more of my potentiality."

Characteristic Features of this Experience

There are a number of characteristics of this experience. The client moves from fearing his inner feelings, and defending himself against them, to letting these feelings *be* and exist in him, as accepted elements of himself. From being out of touch with some aspects of his experience he moves toward a freer inner communication, a greater awareness of what is going on from moment to moment within. A client says, "The real truth of the matter is that I'm *not* the sweet forbearing guy that I try to make out that I am. I get irritated at things. I feel like snapping at people, and I feel like being selfish at times; and I don't know why I should pretend I'm *not* that way."

The client also moves from living by values introjected from others to values which are experienced in himself in the present. From existing only to satisfy the expectations of others, he moves toward being a person in his own right, with feelings, goals and ideas of his own. Thus a young woman says, "I've always tried to be what the others thought I should be, but now I'm wondering whether I shouldn't just see that I am what I am."

Another important element of this experience is that the client moves from being a person driven and compelled by internal and external forces beyond his control, toward being a person who makes responsible choices. One client tells how he has always felt his family was to blame for all of his difficulties, and then he adds, "but now that I understand all they've done, I guess it's up to me." A man who has been in a state hospital for years improves in therapy to a point where he is now facing a most perplexing situation regarding his leaving the hospital. Reflecting both his confusion and his newly born autonomy, he says, "I don't know *what* I'm gonna do, but *I'm* gonna do it."

Another characteristic of this experience is that the client moves from a distrust of the spontaneous and unconscious aspects of himself to a basic trust of his experiencing, and of his organism, as a sound instrument for encountering life. Clients find many ways of expressing this tentative movement into a greater confidence in the deeper aspects of themselves. One man says, "I have a feeling that what I have to do is to leave the vantage points that I have now—from which I look myself over. In a way being *less* conscious, more spontaneous. Take more the position of passenger than driver. See how things go when they're left alone. It's awful kind of scary." At another time he expresses the same feeling when talking about the secret thoughts in himself. "The butterflies are the thoughts closest to the surface. Underneath there's a deeper flow. . . . The deeper flow is like a great school of fish moving under the surface. I see the ones that break through the surface of the water, sitting with my fishing line in hand, trying to find a better tackle. Or better yet, trying to find a way of diving in. That's the scary thing. The image I get is that *I* want to be one of the fish myself." The therapist says, "You want to be down there flowing along, too." This desire to be one with the subterranean and primitive spontaneity within is a real part of the experience I am trying to describe.

Still another element of this experience is the ambivalent and fearful way in which the client moves toward this responsible freedom. It is not an easy thing to have the courage to be, and clients shrink from it at the same time as they move toward it. Thus a young woman who has taken a large step forward in realizing "I can't depend on someone else to *give* me an education. I'll really have to get it by myself," follows this up a moment later by saying, "I have a feeling of strength, and yet I have a feeling of realizing it's so sort of fearful, of fright."

I hope that what I have said thus far conveys some sense of what I mean when I say that clients, in a satisfactory therapeutic relationship, undergo a self-initiated process of learning to be free. This learning is composed of movement from as well as movement toward. From being persons driven by inner forces they do not understand, fearful and distrustful of these deeper feelings and of themselves, living by values they have taken over from others, they move significantly. They move toward being persons who accept and even enjoy their own feelings, who value and trust the deeper layers of their nature, who find strength in being their own uniqueness, who live by values they experience. This learning, this movement, enables them to live as more individuated, more creative, more responsive, and more responsible persons. Clients are, as I have tried to indicate, often sharply aware of such directions in themselves, as they move with fearfulness toward being freely themselves.

The Modern View that Man is Unfree

To some, it must seem strangely out of tune with the modern world to speak, as I have, of learning to be free. The growing opinion today is that man is essentially unfree. He is unfree in a cultural sense. He is all too obviously a pawn of government. He is molded by mass propaganda into being a creature with certain opinions and beliefs, desired and pre-planned by the powers that be. He is the product of his class—lower, middle, or upper—and his values and his behavior are shaped by the class to which he belongs. So it seems increasingly clear from the study of social institutions and influences, that man is simply the creature of his culture and his circumstances, and most decidedly is not free.

At a still deeper level the behavioral sciences have added to this conception of man as unfree. Man is determined in part by his heredity—in his intelligence, his personality type, perhaps even his tendency toward mental aberration. He is above all the product of his conditioning—the inevitable result of the fortuitous events which have "shaped up" his behavior. Many of our most astute behavioral scientists are agreed that this process of conditioning, of "shaping up" the individual's behavior, will not much longer be left to chance, but will be planned. Certainly the behavioral sciences are developing a technology which will enable us to control the individual's behavior to a degree which at the present moment would seem fantastic.

Along with the development of this technology has gone

an underlying philosophy of rigid determinism in the psychological sciences which can perhaps best be illustrated by a brief exchange which I had with Professor B. F. Skinner of Harvard at a recent conference. A paper given by Dr. Skinner led me to direct these remarks to him. "From what I understood Dr. Skinner to say, it is his understanding that though he might have thought *he chose* to come to this meeting, might have thought he had a purpose in giving this speech, such thoughts are really illusory. He actually made certain marks on paper and emitted certain sounds here simply because his genetic makeup and his past environment had operantly conditioned his behavior in such a way that it was rewarding to make these sounds, and that he as a person doesn't enter into this. In fact if I get his thinking correctly, from his strictly scientific point of view, he, as a person, doesn't exist." In his reply Dr. Skinner said that he would not go into the question of whether he had any choice in the matter (presumably because the whole issue is illusory) but stated, "I do accept your characterization of my own presence here." I do not need to labor the point that for Dr. Skinner the concept of "learning to be free" would be quite meaningless.

Thus, though there are opposing voices, the general thrust of the cultural trend throughout both the Western and Communist world is to say that man is not free, that there is no such thing as a free man. We are formed and moved by forces—cultural forces without, and unconscious forces within—which we do not comprehend and which are beyond our control. We will soon be formed more knowingly and more precisely by a scientific technology which will replace the crude way in which we have been molded by partially fortuitous natural events.

The Inadequacy of the "Scientific" View

I have contended, and continue to contend, that this is not the whole picture. The experience I have had with my clients causes me profoundly to disagree with the notion that the individual is no more than a link between a series of complex causes and their inevitable and predetermined effects. When I think of the explanation in which Skinner concurs as to his presence at the conference, I cannot make it apply to human events as I know them. When I try to tell myself, for example, that a Freedom Rider did not choose to expose himself to danger, did not voluntarily risk his life for a right which he valued, and had, as a person, no part in his behavior, my judgment rebels. When I try to tell myself that he behaved in this way, went into a dangerous situation, accepted a brutal beating, served a jail sentence, simply because his genetic constitution and his individual and cultural conditioning caused him to move in certain geographical directions, emit certain sounds when beaten, and further vocalizations when arrested, and that all of these behaviors were emitted because he had been conditioned to find them rewarding—this seems to me a most inadequate and degrading view of man. He becomes a meaningless phenomenon in a world which has no sense.

It is clear, however, that if I object to the concept of man as a meaningless molecule in an equation which he had no part in writing, then I must be willing to define what I mean when I speak of freedom, when I say that I have observed in others, and have experienced in myself, the process of learning to be free. This may seem especially difficult since, as a behavioral scientist, I quite agree with Dr. Skinner in the view that the sequences of cause and effect appear to operate quite as much in the psychological as in the physical world.

What is the Meaning of Freedom?

So what is this freedom of which we speak? In what sense does a client—in what sense can any person—learn to be free? What possible definition of freedom can there be in a modern world?

In the first place, the freedom which I have been trying to describe is essentially an inner thing, something which exists in the living person, quite aside from any of the outward choice of alternatives which we so often think of as constituting freedom. I am speaking of the kind of freedom which Frankl vividly describes in his experience of the concentration camp, when everything—possessions, identity, choice—was taken from the prisoners. But even months and years in such an environment showed only "that everything can be taken from a man but one thing: the last of the human freedoms—to choose one's own attitude in any given set of circumstances, to choose one's own way." It is this inner, subjective, existential freedom which I have observed. It is the realization that "I can live myself, here and now, by my own choice." It is the quality of courage which enables a person to step into the uncertainty of the unknown as he chooses himself. It is the discovery of meaning from within oneself, meaning which comes from listening sensitively and openly to the complexities of what one is experiencing. It is the burden of being responsible for the self one chooses to be. It is the recognition by the person that he is an emerging process not a static end product. The individual who is thus deeply and courageously thinking his own thoughts, becoming his own uniqueness, responsibly choosing himself, may be fortunate in having hundreds of objective outer alternatives from which to choose, or he may be unfortunate in having none, but his freedom exists regardless. So we are first of all speaking of something which exists within the individual, of something phenomenological rather than objective, but none the less to be prized.

A second point in defining this experience of freedom is that it exists not as a contradiction to the picture of the psychological universe as a sequence of cause and effect, but as a complement to such a universe. Freedom, rightly understood, is a fulfillment, by the person, of the ordered sequence of his life. As Martin Buber puts it, "The free man . . . believes in destiny, and believes that it stands in need of him." He moves out voluntarily, freely, responsibly, to play his significant part in a world whose determined events move through him and through his spontaneous choice and will. Again to quote Buber. "He who forgets all that is caused and makes decision out of the depths . . . is a free man, and destiny confronts him as the counterpart of his freedom. It is not his boundary but his fulfillment." This is the answer of the modern philosopher to the prevailing view that man is no more than the sum of his conditioning. Even more convincing than the intellectual answer is the experience of one client after another, as he moves in therapy toward an acceptance of the realities of the world outside and inside himself, and also moves toward becoming a responsible agent in this real world.

We are speaking then, of a freedom which exists in the

subjective person, a freedom which he courageously uses to live his potentialities. We are speaking of a freedom in which the individual chooses to fulfill himself by playing a responsible and voluntary part in bringing about the destined events of his world. This experience of freedom is for my clients a most meaningful development, one which assists them in becoming human, in relating to others, in being a person.

The Facilitation of Inner Freedom

Interestingly enough, we now have a considerable body of knowledge, both clinical and empirical, as to the conditions which, in psychotherapy, foster the process of learning to be free, of becoming one's self. Essentially, we have found, this experience comes about in a close, warm, understanding relationship in which there is freedom *from* such things as threat, and freedom *to* choose and be. Let me spell this out in somewhat more detail.

From the practical and research information currently available, it seems that a growth-facilitating or freedom-promoting relationship contains at least three significant qualities. I would like to describe these in the very briefest fashion, and in everyday language, as we have studied them in psychotherapy.

It has been found that personal change is facilitated when the psychotherapist is what he *is*, when in the relationship with his client he is genuine and without "front" or facade, openly being the feelings and attitudes which at that moment are flowing *in* him. We have coined the term congruence to try to describe this condition. By this we mean that the feelings the therapist is experiencing are available to him, available to his awareness, and he is able to live these feelings, be them, and able to communicate them if appropriate. No one fully achieves this condition, yet the more the therapist is able to listen acceptantly to what is going on within himself, and the more he is able to be the complexity of his feelings, without fear, the higher the degree of his congruence.

To give a commonplace example, each of us senses this quality in people in a variety of ways. One of the things which offends us about radio and TV commercials is that it is often perfectly evident from the tone of voice that the announcer is "putting on," playing a role, saying something he doesn't feel. This is an example of incongruence. On the other hand each of us knows individuals whom we somehow trust because we sense that they are being what they are, that we are dealing with the person himself, not with a polite or professional front. It is this quality of congruence which we sense that research has found to be associated with successful therapy. The more genuine and congruent the therapist in the relationship, the more probability there is that change in personality in the client will occur.

Now the second condition. When the therapist is experiencing a warm, positive and acceptant attitude toward what *is* in the client, this facilitates change. It involves the therapist's genuine willingness for the client to be whatever feeling is going on in him at that moment—fear, confusion, pain, pride, anger, hatred, love, courage, or awe. It means that the therapist cares for the client, in a non-possessive way. It means that he prizes the client in a total rather than a conditional way. By this I mean that he does not simply accept the client when he is behaving in certain ways, and disapprove of him when he behaves in other ways. It means an outgoing positive feeling without reservations, without evaluations. The term we have come to use for this is unconditional positive regard. Again research studies show that the more this attitude is experienced by the therapist, the more likelihood there is that therapy will be successful.

The third condition we may call emphatic understanding. When the therapist is sensing the feelings and personal meanings which the client is experiencing in each moment, when he can perceive these from "inside," as they seem to the client, and when he can successfully communicate something of that understanding to his client, then this third condition is fulfilled.

I suspect each of us has discovered that this kind of understanding is extremely rare. We neither receive it nor offer it with any great frequency. Instead we offer another type of understanding which is very different. "I understand what is wrong with you"; "I understand what makes you act that way"; or "I too have experienced your trouble and I reacted very differently"; these are the types of understanding which we usually offer and receive, an evaluative understanding from the outside. But when someone understands how it feels and seems to be *me*, without wanting to analyze me or judge me, then I can blossom and grow in that climate. And research bears out this common observation. When the therapist can grasp the moment-to-moment experiencing occurring in the inner world of the client as the client sees it and feels it, without losing the separateness of his own identity in this emphatic process, then change is likely to occur.

Studies with a variety of clients show that when these three conditions occur in the therapist, and when they are to some degree perceived by the client, therapeutic movement ensues, the client finds himself painfully but definitely learning and growing, and both he and the therapist regard the outcome as successful. It seems from our studies that it is attitudes such as these rather than the therapist's technical knowledge and skill, which are primarily responsible for therapeutic change.

The Dynamics of a Developing Inner Freedom

You may well ask, "But why does a person who is seeking help find himself changing in a relationship which contains these elements? Why does this initiate a process of learning to be free, of becoming what he is, of choice and inner development?" Let me try very briefly to answer such questions.

The reactions of the client who experiences for a time the kind of therapeutic relationship which I have described are a reciprocal of the therapist's attitudes. In the first place, as he finds someone else listening acceptantly to his feelings, he little by little becomes able to listen to himself. He begins to receive the communications from within himself—to realize that he *is* angry, to recognize when he is frightened, even to realize when he is feeling courageous. As he becomes more open to what is going on within him he becomes able to listen to feelings which have seemed to him so terrible, or so disorganizing, or so unique, or so personal, that he has never been able to recognize their existence in himself.

While he is learning to listen to himself he also becomes more acceptant of himself. As he expresses more and more of the hidden aspects of himself, he finds the therapist showing a consistent and unconditional positive regard for him and his feelings. Slowly he moves toward taking the same attitude toward himself, accepting himself as he is, respecting and caring for himself as a person, being responsible for himself as he is, and therefore ready to move forward in the process of being free.

And finally as he listens more accurately to the feelings within, and becomes less evaluative and more acceptant toward himself, he also moves toward being more real. He finds it possible to move out from behind the facade he has used, to drop his defensive behaviors, and more openly to be what he truly is. As these changes occur, as he becomes more self-aware, more self-acceptant, more self-expressive, less defensive and more open, he finds that he is at last free to change and grow and move in the directions natural to the human organism. He can make imperfect choices—and then correct them. He recognizes that he can choose to be hurtful or constructive, self-aggrandizing or committed to the welfare of the group, and when these choices can be freely made, he tends to move in the socially constructive direction.

This is a brief picture of what we have seen time and again in the experience of psychotherapy.

A Hope for Education

It is such experiences in individual and group psychotherapy which lead us to believe that we have here an important dynamic for modern education. We may have here the essential core of a process by which we might facilitate the production, through our educational system, of persons who will be adaptive and creative, able to make responsible choices, open to the kaleidoscopic changes in their world, worthy citizens of a fantastically expanding universe. It seems at least a possibility that in our schools and colleges, in our professional schools and universities, individuals could learn to be free.

The Current Trend Toward Hard-Headed Conformity

I say this in full recognition of the fact that the current trend in education is away from freedom. There are tremendous pressures today—cultural and political—for conformity, docility, and rigidity. The demand is for technically trained students who can beat the Russians, and none of this nonsense about education which might improve our interpersonal relationships! The demand is for hard-headedness, for training of the intellect only, for scientific proficiency. We want inventiveness in developing better "hardware," but creativity in a larger sense tends to be suspect. Personal feelings, free choice, uniqueness—these have little or no place in the classroom. One may observe an elementary school classroom for hours without recording one instance of individual creativity of free choice, except when the teacher's back is turned. And at the college level we know that the major effect of a college education on the values of the student is to "shape up" the individual for more comfortable membership in the ranks of college alumni.

I am, therefore, quite aware that for the general public and for most educators the goal of learning to be free is not an aim they would select, nor toward which they are actually moving. Yet if a civilized culture is to survive, and if the individuals in that culture are to be worth saving, it appears to me to be an essential goal of education. So I would like to explore something of what it means when we take seriously the learnings from psychotherapy and endeavor to apply them to the field of education in order to foster the development of persons who are inwardly free.

Some Efforts to Permit Freedom in Education

The past has not been devoid of such experimentation, and there has accumulated a considerable body of practical experience, and a smaller body of empirical knowledge regarding education which has inward freedom as one of its primary goals. I will comment very briefly on four of the practical efforts along this line.

August Aichorn, many years ago, carried on a radical experiment in the reeducation of delinquents. He permitted them freedom, within the institutional setting, to conduct themselves as they desired in the group in which he was the leader. After a period of chaos which I am sure few of us could bear, these youths gradually chose a social and disciplined and cooperative life as something they *preferred.* They learned, through experience in an accepting relationship, that they desired responsible freedom and self-imposed limits rather than the chaos of license and aggression.

Another radical experiment was that conducted by A. S.

Neill in his school, Summerhill. Started forty years ago, this school has become a current focus of great interest because of A. S. Neill's book *Summerhill* telling of the experiences of his pupils and himself in this school. This is a book which is well worth a thoughtful reading by every educator. Neill's sincerity and genuineness, his faith in the potential of each individual, his firm respect for each child and for himself, shines through its pages. As in the case of Aichorn, few of us would have the courage to trust the individual, and his natural desire to learn, as completely as does Neill. Yet he has given us a challenging laboratory example of what it means to provide a setting in which children can learn to be free. Even the cautious report of the Ministry of Education makes it clear that the students develop a zest for living, a spontaneous courtesy, as well as initiative, responsibility, and integrity. They conclude, "a piece of fascinating and valuable educational research is going on here which it would do all educationists good to see."

The core of the progressive education movement, now so frequently derided, was another attempt to help individuals to learn to be free. That its fundamental philosophy frequently became debased into turning education into a sugar-coated pill should not obscure its true aims, nor its effective results when it was true to its own philosophical base.

Still another type of experiment along this line is evident in the work being done in student-centered teaching. Here much of the work has been done in university classes and in intensive workshops for professional persons. The aims of such an approach have been summarized in the following terms:

> The goal . . . . is to assist students to become individuals
> who are able to take self-initiated action and to be responsible for those actions;
> who are capable of intelligent choice and self-direction;
> who are critical learners, able to evaluate the contributions made by others;
> who have acquired knowledge relevant to the solution of problems;
> who, even more importantly, are able to adapt flexibly and intelligently to new problem situations.

who have internalized an adaptive mode of approach to problems, utilizing all pertinent experience freely and creatively;

who are able to cooperate effectively with others in these various activities;

who work, not for approval of others, but in terms of their own socialized purposes.

Some Conditions Which Facilitate "Learning to be Free"

If we review all of these streams of effort, it seems possible to abstract from the various experiences, and from the pertinent research, those conditions which appear to be essential if we are to facilitate in students this quality of inward freedom. I should like to describe these conditions as I see them.

Confronting a Problem

In the first place, if this self-initiated learning is to occur, it seems essential that the individual be in contact with, be faced by, a real problem. Success in facilitating such learning often seems directly related to this factor. Professional persons who come together in a workshop, because of a concern with problems they are facing, are a good example. Almost invariably, when they are given the facilitating climate I will describe, they at first resist the notion of being responsible for their own learning, and then seize upon this as an opportunity, and use it far beyond their expectations. On the other hand, students in a required course expect to remain passive, and may find themselves extremely perplexed and frustrated at being given freedom. "Freedom to do what?" is their quite understandable question.

So it seems reasonably clear that for learning of the sort we are discussing it is necessary that the student, of whatever level, be confronted by issues which have meaning and relevance for him. In our culture we tend to try to insulate the student from any and all of the real problems of life, and this constitutes a difficulty. It appears that if we desire to have students learn to be free and responsible individuals, then we must be willing for them to confront life, to face problems. Whether we are speaking of the inability of the small child to make change, or the problem of his older brother in constructing a hi-fi set, or the problem of the college student and adult in formulating his views on international policy, or dealing effectively with his interpersonal relationships, some real confrontation by a problem seems a necessary condition for this type of learning.

A Trust in the Human Organism

I come now to those conditions which are essentially dependent upon the teacher. It is clear from the experience of Aichorn, Neill, or the many individuals who have tried a student-centered approach to teaching, that one of the requisites for the teacher who would facilitate this type of learning is a profound trust in the human organism. If we distrust the human being, then we *must* cram him with information of our own choosing, lest he go his own mistaken way. But if we trust the capacity of the human individual for developing his own potentiality, then we can permit him the opportunity to choose his own way in his learning. Hence it is evident that the kind of learning I am discussing would be possible only for a teacher who holds a somewhat confident view of man.

Realness in the Teacher

Another element of the teacher's functioning which stands out is his sincerity, his realness, his absence of a facade. He can be a real person in his relationship with his students. He can be angry. He can also be sensitive and sympathetic. Because he accepts his feelings as his own, he has no need to impose them on his students. He can dislike a student product without implying that it is objectively bad or that the student is bad. It is simply true that he, as a person, dislikes it. Thus he is a *person* to his students, not a faceless embodiment of a curricular requirement, not a sterile tube through which knowledge is passed from one generation to the next.

Acceptance

Another attitude which stands out in the work of those who have been successful in promoting this type of learning is a prizing of the student, a prizing of his feelings and opinions. The teacher values the individual student as having worth, and this prizing extends to each and all the facets of this individual. Such a teacher can be fully acceptant of the fear and hesitation of the student as he approaches a new problem, as well as of the satisfaction he feels in achievement. If the teacher can accept the student's occasional apathy, his desire to explore by-roads of knowledge, as well as his disciplined efforts to achieve major goals, he will promote this type of learning. If he can accept personal feelings which both disturb and promote learning—rivalry with a sibling, hatred of authority, concern about personal adequacy—then he is certainly such a teacher. I trust I am making it clear that this means an acceptance of the whole student by the teacher—a prizing of him as an imperfect human being with many feelings, many potentialities. This prizing or acceptance is an operational expression of the teacher's essential confidence in the capacity of the human organism.

Empathy

Still another element in the teacher's attitude is his ability to understand the student's reactions from the inside, an empathic awareness of the way the process of education and learning seems to the student. This is a kind of understanding almost never exhibited in the classroom; yet when the teacher *is* empathic, it adds an extremely potent aspect to the classroom climate. When a child says, in a discouraged voice, "I can't do this," that teacher is most helpful who naturally and spontaneously responds, "You're just hopeless that you can ever learn it, aren't you?" The usual denial of the child's feeling by a teacher who says, "Oh but I'm *sure* you can do it" is not nearly so helpful.

Providing Resources

These then are the essential attitudes of the teacher who facilitates a learning to be free. There is one other function performed by such a teacher which is very important. It is the provision of resources. Instead of organizing lesson plans and lectures, such a teacher concentrates on providing all kinds of relevant raw material for use by the students, together with clearly indicated channels by which the student can avail himself of these resources. I am thinking not only of the usual academic resources—books, workspace, tools, maps, movies, recordings, and the like. I am also thinking of human resources—persons who might contribute to the knowledge of the student. Most important in this respect is the teacher himself as a resource. He makes himself and his special knowledge and experience clearly available to the students, but he does not impose himself on them. He outlines the particular ways in which he feels he is most competent, and they can call on him for anything he is able to give, but this is an offer of himself as a resource, and the degree to which he is used is up to the students.

What the Teacher Does Not Do

The teacher thus concentrates on creating a facilitative climate and upon providing resources. He may also help to put students in contact with meaningful problems. But he does not set lesson tasks. He does not assign readings. He does not lecture or expound (unless requested to). He does not evaluate and criticize unless the student wishes his judgment on a product. He does not give examinations. He does not set grades. Perhaps this will make it clear that such a teacher is not simply giving lip service to a different approach to learning. He is actually, operationally, giving his students the opportunity to learn to be responsibly free.

The Process of Learning to be Free

When the teacher establishes an attitudinal climate of the sort I have described, when he makes available resources which are relevant to problems which confront the student, then a typical process ensues.

Initial Frustration

First, for students who have been taught by more conventional means, there is a period of tension, frustration, disappointment, disbelief. Students turn in such statements as "I felt completely frustrated by the class procedure." "I felt totally inadequate to take part in this kind of thing." "The class seems to be lacking in planning and direction." "I keep wishing the *course* would start."

One mature participant observer describes the way one group struggled with the prospect of freedom after an initial session in which opportunities and resources were described.

> Thereafter followed four hard, frustrating sessions. During this period, the class didn't seem to get anywhere. Students spoke at random, saying whatever came into their heads. It all seemed chaotic, aimless, a waste of time. A student would bring up some aspect of the subject; and the next student, completely disregarding the first, would take the group away in another direction; and a third, completely disregarding the first two, would start fresh on something else altogether. At times there were faint efforts at a cohesive discussion, but for the most part the classroom proceedings seemed to lack continuity and direction. The instructor received every contribution with attention and regard. He did not find any student's contribution in order or out of order.
>
> The class was not prepared for such a totally unstructured approach. They did not know how to proceed. In their perplexity and frustration, they demanded that the teacher play the role assigned to him by custom and tradition; that he set forth for us in authoritative language what was right and wrong, what was good and bad.

This is a good description of the bafflement and chaos which is an almost inevitable initial phase of learning to be free.

Individual Initiative and Work

Gradually students come to various realizations. It dawns on them that this is not a gimmick, but that they are really unfettered; that there is little point in impressing the professor, since the student will evaluate his own work; that they can learn what they please; that they can express, in class, the way they really feel; that issues can be discussed in class which are real to them, not simply the issues set forth in a text. When these elements are recognized, there is a vital and almost awe-inspiring release of energy. One student reads as she has never read before—two books a week in the subject, and hopes this "will never end." Others undertake projects of writing, experimentation, work in a clinic or laboratory, with a new zest. The report of one student is typical of many, and is worth quoting at some length.

> I feel that I want to share my joy with you in relation to the paper that I gave you earlier today—it is what I call "my first *real* learning experience". . . .
>
> I took a few minutes after I finished typing my paper to think what had made this learning experience so different from the many others which I have had. These are my reactions, sketched briefly:
>
> Based on *real* need—not superficial topic. . . .
> Reading was done to satisfy my need, not merely to collect material to fit topic and sound good. . . .
>
> I found that I had to scrap my original approach toward writing a paper when I realized that it did not have to sound good or conform to a prescribed pattern. I jotted down my usual idea of a good outline for a paper only to find that it was not geared to my need at all, and I turned to writing about things of significance to me and then made an outline of what I had written.
>
> One of the most "shocking" parts of this experience, as I have related to you one day, was the fact that I did not have to do this and yet I wanted to be working on it all the time and rushed through assigned requirements in other courses to devote time to this.
>
> I wrote an annotated bibliography for the first time in my life because I wanted to have information regarding this material I had read, for future reference. . . .
>
> There was no feeling of drudgery about this paper—I found myself saying, "I'm going over to the library to work on my paper for a while" instead of, "Oh, I sup-

> pose. I've got to plow through some more books tonight or I'll never get that paper done on time." The lack of external pressure made this experience one of the most enjoyable things I have ever done. Basically, through experience, it has changed my whole aproach to teaching. . . .

It is clear that this student is discovering what it means to be autonomous, what it means to be creative, what it means to put forth disciplined effort to reach one's own goals, what it means to be a responsible free person, and most important, is appreciating the satisfactions which come from these experiences.

Personal Closeness

Another element which is a common part of the process is that the group develops a respect and liking for each other as individuals, as they emerge in the group discussions. A teacher trying this approach writes, "In this second group, also, I found that the students had developed a personal closeness, so that at the end of the semester they talked of having annual reunions. They said that somehow or other they wanted to keep this experience alive and not lose one another."

Individual Change

As the learning continues, personal changes take place in the direction of greater freedom and spontaneity. Here is another report by a participant observer.

> In the course of this process, I saw hard, inflexible, dogmatic persons, in the brief period of several weeks, change in front of my eyes and become sympathetic, understanding and to a marked degree non-judgmental. I saw neurotic, compulsive persons ease up and become more accepting of themselves and others. In one instance, a student who particularly impressed me by his change, told me when I mentioned this: "It is true. I feel less rigid, more open to the world. And I like myself better for it. I don't believe I ever learned so much anywhere."

> I saw shy persons become less shy and aggressive persons more sensitive and moderate.

A more personal statement of this kind of change is given by a student at the end of the course.

> Your way of being with us is a revelation to me. In your class I feel important, mature and capable of doing things on my own. I want to think for myself and this need cannot be accomplished through text books and lectures alone, but through living. I think you see me as a person with real feelings and needs, an individual. What I say and do are significant expressions from me, and you recognize this. You follow no plan, yet I'm learning. Since the term began I seem to feel more alive, more real to myself. I enjoy being alone as well as with other people. My relationships with children and other adults are becoming more emotional and involved. Eating an orange last week, I peeled the skin off each separate orange section and liked it better with the transparent shell off. It was juicier and fresher tasting that way. I began to think, that's how I feel sometimes, without a transparent wall around me, really communicating my feelings. I feel that I'm growing, how much, I don't know. I'm thinking, considering, pondering and learning.

Throughout this description of the learning process in such a climate, I am sure you will have observed the many similarities to the process involved in psychotherapy. Outstanding is the way in which the student begins to rely on his own values as he experiences them, rather than upon the values imposed on him by others. It is also clear that the student is closer to his own feelings, trusts them more, trusts himself more. He is not so afraid of his own spontaneity, not so afraid of change. He is, in short, learning what it means to be free.

Effect Upon the Instructor

I believe the story of this kind of classroom experience is incomplete without some mention of the effect upon the instructor when he has been the agent for the release of such self-initiated learning. One such teacher says, "To say that I am overwhelmed by what happened only faintly reflects my feelings. I have taught for many years but I . . . never have found in the classroom so much of the whole person coming forth, so deeply involved, so deeply stirred . . . I can only . . . say that I am grateful and I am also humbled by the experience."

Another report as follows: "Rogers has said that relationships conducted on these assumptions mean 'turning present day education upside down.' I have found this to be true as I have tried to implement this way of living with students. The experiences I have had have plunged me into relationships which have been significant and challenging beyond compare for me. They have inspired me and stimulated me and left me, at times, shaken and awed with their consequences for both me and the students. They have led me to the fact of what I can only call . . . the tragedy of education in our time—student after student who reports this to be his first experience with total trust, with freedom to be and to move in ways most consistent for the enhancement and maintenance of a core of dignity which somehow has survived humiliation, distortion and corrosive cynicism."

Research Corroboration

Empirical investigations of this sort of teaching I have described are neither large in number, nor noteworthy for their research sophistication. Yet they bear out the student report quoted above in indicating that improvement in personal psychological maturity is significantly greater in student-centered classes than in conventional ones. There is also a greater amount of self-initiated extra-curricular learning, and evidence of greater creativity and self-responsibility. As to the factual and curricular learning, this seems roughly equal to that achieved in conventional classes. Some studies report slightly more, others slightly less. The fairest summary seems to be that if we are solely concerned with the teaching of teacher-selected content material, this approach is probably no better and no worse than the ordinary class. If we are concerned with the development of the person, with initiative, originality and responsibility, such an approach produces greater changes.

Summary

Let me retrace briefly the pathway I have followed in this talk. I have pointed out that though freedom is a concept not at all acceptable to our modern intellectual culture, it is something which is undeniably experienced by clients in individual and group therapy. These individuals, when permitted to live, even for a small fraction of their time, in the special psychological climate of the therapeutic relationship, achieve a spontaneous, existential, and creative inner freedom. I have tried to show that the freedom they achieve is definable in ways which complement rather than contradict the current scientific view of man's behavior.

When students are permitted to be in contact with real problems; when resources—both human and technical—are made psychologically available by the teacher; when the teacher is a real person in his relationships with students, and feels an acceptance of and an empathy toward his students; then an exciting kind of learning occurs. Students go through a frustrating but rewarding process in which gradually responsible initiative, creativity, and inner freedom are released. The kind of personal and intellectual change which comes about has many parallels with the changes which occur in psychotherapy. The nature of these changes has to some extent been investigated empirically.

In closing, I would like to point out that for the most part modern culture does not, operationally, want persons to be free, in spite of many ideological statements to the contrary. Both of the two main streams of modern life—Western and Communist—are extremely fearful of and ambivalent about any process which leads to inner freedom. Nevertheless it is my personal conviction that individual rigidity and constricted learning are the surest roads to world catastrophe. It seems clear that if we prefer to develop flexible, adaptive, creative individuals, we have a beginning knowledge as to how this may be done. We know how to establish, in an educational situation, the conditions and the psychological climate which initiate a process of learning to be free.

III

> I reject any organized pretense to an objective knowledge of man. I know only what I sweat from my own personal struggle to stay alive. Psychotherapy is not a professional routine. It is a personal venture. The client is "like me." I reject any professional boundary between us. I "make it" as a person or I fail.
>
> Richard Johnson (psychotherapist)

Becoming a person
Freedom to be
Courage to be
Learning to be free

All these labels mean to me what I call "getting together with myself." When I do this, I feel more free and, in my own words, "more like a person." Sometimes I call this "looking through my own windows instead of someone else's"—seeing what *I* see, not what someone else thinks that I should see. I don't always do this, but now I am moving clearly in the direction of being more myself.

This is very different from when I felt unfree and didn't know *how* I was unfree. It seemed that I "had done something wrong" without the faintest notion what the wrongness was. I felt accused, but who was accusing me? Myself? How could myself accuse me when I didn't know what I was accused of?

I had lost something. That I knew, and that was all I knew. How could I find it when I didn't know what it was or where to look for it and other people were no help to me? They said that I shouldn't be restless, shouldn't be dissatisfied, that I should be happy with what I had, and that's the way life *is* and I should put up with it as everyone else did. These were the things that I had been told when I was eighteen, but when I was eighteen I wasn't so confused: I knew what *I* wanted, and I didn't feel guilty. It was also clear to me then that life is the way it is because everyone says "That's the way it is" and goes along with it.

All that clarity had got so lost later on that when I did what seemed right and good *to me*, I felt "irresponsible," but I was in a complete fog about *why*. Getting out of that certainly was "a confused and groping movement." Sometimes, though, there was a sudden flash of seeing clearly—a brief, bright instant—but then that got lost and I had to grope my way back to something that I knew was there although I couldn't remember what it was. Sometimes it was like being for an instant on a mountain top, taking in the whole scene—all the different facets of the mountain and all the different views around it. While I was there, I understood everything. But then I wasn't there any more. I was a half-inch high person at the bottom of the mountain that I had to find my way back to the top of. Sometimes it seemed that I was walking underground for months and miles, then suddenly I was on top and everything could be seen. But after only a few steps I was underground again, dropped like an elevator going down.

My fish image was that I *was* a fish, struggling to swim against the stream. I wasn't able to do that yet. I was held where I was, which gave a chance to learn how to swim against the stream (go the way that I wanted to go) by two shadowy figures on a bridge, each of them holding a line which I was hooked to like a fish. This kept me from being swept away by the swiftly-moving stream. I knew that the two figures were the doctor and Aldous Huxley. Neither of them understood everything, but each of them understood enough to be helpful. When I got worse physically, the doctor got me back to some degree of steadiness. When, through experimenting with my mind, I got into something that neither the doctor nor I understood, so I was afraid to go on with it, I wrote to Huxley and he explained it. These two men kept me from being swept away while I thrashed around, learning how to swim against the stream. This was a persistent image that I lived with for about a year.

These men were also people to whom I could tell anything and they wouldn't "call the cops." This meant to me that they wouldn't call the men to lock me up. I thought of being "locked up" as being in a madhouse, but it wasn't a mental hospital that I was afraid of, although I didn't know what the "madhouse" was. It meant being pushed back into what I was struggling to get out of. Other people tended very much to do this to me, so I lived more and more alone and when, at last, I was just barely able to travel I went to a place where I knew no one, and kept myself alone, so that I could get together with myself. I wanted desperately to be with someone who understood more than I did about what I was trying to understand, but since I couldn't do that, I could at least remove myself from people who were confusing me.

After all that, some of my present knowings seem small and perhaps ridiculous, but I know now that they are *not* "unimportant." I am living near the beach in an apartment which has an outside deck with a railing. When my son was here, he started to throw his damp swimming trunks and towel over the railing, then said, "The management probably wouldn't like that—I can see why." I agreed, and his statement was accurate, but whose seeing was the seeing *why?* When I agreed, in my mind there was an image of the uncluttered railing as *desirable,* something that *I* liked. But a few days later when I walked past a building with a railing draped with swim suits and towels, I knew that I *liked* this, that to me it looked gay and human, alive with an *activity* of people. Then I knew how much I miss seeing clothes on lines blowing in the wind, people working untidily in gardens, sweeping sidewalks, dashing out of houses half-dressed to do something that should be done right now, or a woman drying her hair in the sun. When I look out on the tidy street with no sign that anyone lives behind the curtains in the windows of the houses, it seems so lifeless. The *alwaysness* of this tidiness tires me the way that hunger does; something is missing from my intake. If no one else feels as I do, this still *is* the way that *I* feel, and when I think that I don't, I am not together with myself.

If I could deceive myself completely by accepting other people's values, then there might be an argument for giving up and letting other people tell me what to do. But my inner valuing does not cease: it just gets buried to my knowing and is forever in conflict with the values that I have accepted from outside. When I hadn't noticed my own valuing of the street on which I live, there was nothing that I could do about it but be irked without knowing why, and feel that I must be ungrateful because "what I have is good" and still I am unhappy. Now that I've noticed, I feel happy. The conflict *in me* has been removed. Having accepted myself, I can accept other things too, in a way that is very different from "making the best of it." It is the way I lived from age 12 to 16, when I wanted to quit school but the law wouldn't let me. So, I lived with what was around me, including school, until the time when I could leave. The circumstances are different now, but the feeling is the same. I don't feel trapped. I don't feel that something has been done to me (victimized). And I don't feel guilty.

There seem to be two levels of guiltiness. The deeper one is when I am not true to myself. My deeper self, then, seems to be reproaching me. The other guilt comes when I do what *is* in accord with me, but without seeing clearly (sometimes not seeing at all) the distinction between what I want and what others say that I should want. When I do what is in accord with me without this clear seeing, I feel "wrong" or "bad" and that I must be somewhat unsane or disreputable to like what I do. The introjected values are like a monitor saying to my own responses, "You mustn't do that!" When I act in accord with me *and* know clearly what I am doing, then I am freed of both guilts at once: myself no longer reproaches me (it expresses content by a feeling of ease and innocence), and the reproaches of other people seem to have nothing to do with *me*. Which of course they don't.

Carl Rogers quotes a client who says, "The real truth of the matter is that I'm *not* the sweet forbearing guy that I try to make out that I am. I get irritated at times. I feel like snapping at people, and I feel like being selfish at times, and I don't know why I should pretend that I'm *not* that way." This statement could seem to mean that it's good to snap at people. I know some people who use Rogers as their authority for popping off with the first thing that comes into their head. That is not what he means, but what follows is what *I* mean:

The place I have to begin is, it's no good *pretending* that I'm *not* that way. When I notice that I'm pretending, and remove myself from that dishonesty, then I am free to notice that I feel like snapping. (If I think that I am not a snappy person and my mind is on that, how can I notice that I feel like snapping? This seems to me simple mechanics.) But if I notice that I feel like snapping and snap, as the snapping people do, then I have not noticed myself in a way that brings about a *basic* change.

My noticing that I want to snap has to be simple acceptance of the fact, without opinion. If I think that it is good to snap, I'm in the same fix that I am if I think that it is bad to snap. If my noticing is that I want to snap *at* or *about*, this is still the wrong noticing. If I feel justified, I'm in the wrong place, too. I must go more deeply inward, and notice simply the feeling of my irritation.

When I notice my irritation in this *inward* way, something changes. I don't know anything about brain circuits, but I must be able to use a switch in some way, because when I have done this noticing, even if I say the same words that I might have said otherwise, they don't *sound* the same, and the sound is part of the message. The "switch" seems to be the same one that I use with someone whom I love very much, when I wish to hurt and at the same time my love is greater than my wish to hurt. I pause, stopping the expression of hurt that was going to come from me, and then my love comes through. This is not the same as *repressing* my hurt. In the pause, I do not review things and think what I am going to say. It is more like a traffic cop putting up his hand to stop the traffic in one way so that the traffic the other way can come through. I *choose* to let my love come through, and this is my own choosing, having nothing to do with commandments from the Bible or anywhere else. My love comes from me and I like it. That is all.

This seems to me to be the same pause that I can make in other circumstances. When a woman annoys me by (as it seems to me) always exaggerating her feelings, and then says dramatically "I'm afraid that if I start crying I'll never stop!" I could easily make a cutting remark. When I noticed my annoyance first, what I actually *said* was, "I think you don't need to worry about that. I think it's like trying to hold your breath—a mechanism takes over and you have to stop." That could have been said with a bite, but it wasn't. By my noticing *first*, without pretending to myself that I was not annoyed, or that I was sorry for this woman (which at the moment I was not, although at some other times I am), the words came out matter-of-factly, with a kind of reassurance in them, along with permission to cry. *Then* I noticed that at a deeper level this was the way that I really *felt*. In spite of my surface annoyance, I *wanted* to reassure her. My human caring came through, and I knew that even if she was putting on an act, even if she behaved in this way to get attention, the fear was there, and what I said (to my own surprise) was directed to her fear. When I was annoyed, her fear was not a part of my awareness. It *had been*, but when I became annoyed I lost touch with that completely.

I don't quote that remark as the best that could have been said, but it does seem to me clearly *better* than to have spoken to her in annoyance or to have kept my annoyance bottled up.

At another time, at a noisy party, when for a moment this woman and I were in a spot of isolation from the crowd, she said, "It's so awful to be sick," referring to her physical sickness. I said, "So awful to be unable to get well," which it seemed to me was what she was feeling. She said, ". . . unable to get well"—so thoughtfully, as though she had been understood and this was helping her to understand herself. I felt very close to her.

My spontaneous remarks are always a surprise to me. There is no copying of others, and not even a copying of me by me (saying something that I have said or thought before.) I like what I have said—feel good about it in a quiet way—and I like the responsiveness that I feel from the other person.

At another party, with people talking in a group, this same woman said, with obvious fear, that she was taking cortisone and people kept telling her all the awful things it had done to their friends. I get tired of this woman throwing her fears into every conversation, no matter what it is about, and there were others present whom I *did* want to hear from. I didn't pause and notice. I said, "I took it for five years. Don't worry about what people tell you." She said, "Oh, I don't!" and I was sad, because it was what I had said and the way I said it that produced that lie, and there was distance between us.

When I told her, "I think you don't need to worry about that" (her fear that her crying would never stop) and "So awful to be unable to get well," there was a non-cold matter-of-factness about it that seems not to foul things up. I have lived this way with Hawaiians, Navajos and Hopis easily because that was so much their way. It was easy for me to respond to their way of responding to me. It is much easier to go along with the way that others are going than it is to buck it. A young woman with a very large and beautiful home and everything that goes with it came home from lunch one day and told me, "I hate myself! They all talk about what they *have*, and then I talk about what *I* have, when (waving her arm to include everything) *I* don't give a damn about *any* of it!" That's the problem. At any rate, it's my problem in my own society where I find it difficult to achieve what I like except with a very few other people or by myself. I have been trying to extend this, to go beyond where it comes easy. But I cannot *make* myself be non-cold matter-of-fact (with caring). When I try for matter-of-factness directly, there is a coldness about it and rigidity, instead of the soft-firmness that I like. When I try to soften and warm the cold, it becomes icky. The only way that I am able to arrive at the warm/soft/firm/clear matter-of-factness that I like is by indirection, by going inward first. Then, it happens of itself. This is what "non-directive" means to me: I have non-directed *myself*. Then, I *am* non-directive with the other person. A change has taken place in me.

Children often speak in this way if they are permitted to, conveying with it all the variety of their inwardness—their love, curiosity, laughter, sadness, caring. "I love you, Mum. I won't when you're old and rumply, but I do now." So matter-of-fact. "I *would* like to see his skeleton." "The way that Dad looks at me is like an iron bean in my heart." "I don't like cities. Every time you fall down, you hit the pavement." "Are you going to stay the whole two weeks?" "You have a funny face." "I would like to *see* your bones break." "Don't you hope that you'll get well before you die?" Statements and questions about things as they seem to me, not what I *should* say, but simply what *I* am thinking before I have been told that I shouldn't think what I do. It isn't flat, and it is so very *real*, so truly honest, when I have not yet learned that I must lie and pretend. I have learned a lot from children, and when I go along with them in *this* way of theirs, I am a better person. When we reinforce each other in this fashion, there is the beginning of a better world. Sometimes young people still retain this matter-of-factness with each other. The young neighbor whom I don't know very well knocks on my door and says, "Would you turn your hi-fi down?" The young man visiting me, a stranger to the neighbor, says easily, "Like now," with a wave of the hand that says, "I get it." Understanding and acceptance. No fuss.

When I *notice* that I am feeling selfish, then I notice also that I *need to be* selfish at this time, and the clear statement of my need comes out matter-of-factly, without appeal or demand. There are no implications. Of course, sometimes when I notice that I am being selfish, I see that I don't need to be, that it was all very silly and I can easily give it up. The noticing in the way that I have attempted to describe seems to arrive at the truth of the moment, undisturbed by notions of any kind. The truth of the moment is free, not bound by anything. To me, this is spontaneity, which includes humor, too, but *this* humor is rarely possible to convey to someone else in an anecdote, because it is so much a part of the unique circumstances of the moment that all the circumstances have to be described; and then the humor is lost because it is the coming together of everything in one moment that is funny. This is a bubbly way to live—I mean the kind of bubbles that come up through soda water. When we strive for bubbles—which seems to happen often in my own society—it seems to me more like what comes out of the top of a percolator under full steam.

One client "tells how he has always felt his family was to blame for all of his difficulties, and then adds, 'But now that I understand all they've done, I guess it's up to me.'" In me, the knowing "It's up to me" comes about in this way: When I understand "all they've done" (to me), I also understand "so this must have been done to them." Through knowing the way that it came about in me, I know how it must have come about in them. Then the whole thing goes back and back, each generation having been "done to" by the preceding one. There is no one to blame, and all that I can do is get to work on myself *now* and break the chain, so that I won't do unto others what has been done to me. It is not enough that I simply *know* this. I can *know* without any action taking place either within me or outside me. (The one follows upon the other.) I can feel superior to someone who doesn't *know* what I know, without being one whit a better person for my knowing. I have to *undo* what has been done to me, by *doing* in another way. That's where the scary part comes in, in my experience: I can find it comforting to *know*, and almost terrifying to *do*. But it is in the doing that I change, and in my changing, my relations with other people change too.

When I moved to a town a thousand miles away where I knew no one, the doctor who had been recommended to me by his training would not cooperate with the regime which the previous doctor and I had found—through painful trial and many errors—worked for me. I decided that it would be better to leave this doctor. The only other specialist in what I needed had moved up a canyon and given up practice, and I didn't know anything at all about the other doctors. I decided that I would have to get along without one. This went well for a while, but then I hit a phase when I got worse in a way that I knew meant that the dosage of cortisone should be adjusted. I couldn't tell whether I should take more or less. I wanted to telephone the previous doctor for help. I didn't, and I felt noble about not calling him. This was the first time that I latched onto what has since proved to be a fact of my life, that when I feel noble about not doing something, it is not *myself* who is doing it. I got worse and worse and felt nobler and nobler. I thought that *I* was not calling the doctor, and in the objective or physical sense this was true: this person, this body, did not make the phone call. But the inside world is not so simple as that. It *could* be, but it has an enormous capacity for getting mixed up because other people have got into me through their directives. I had a dream which made this clear to me. It is really beautiful, the way that something inside me goes on seeing clearly, even when what I call my "conscious" mind is so mixed-up that it is hell I live in.

In my dream, I came into a room where a girl sat rigidly on a bench with her eyes tight shut, in pain from sunlight that was glaring through a window in a band across her eyes. I thought how silly she was to sit there in torment, when all that she had to do was move and then she would be relieved and her eyes could open. I (feeling very superior about taking action), went to the window and let down the venetian blind, to free her. There were many people sitting around a refectory table absorbed in their conversation with each other. They didn't notice the girl or me, or anything we did. They had no interest in us.

When I woke up, I lived with that dream until I knew that

the foolish girl was *me*—or that part of myself which is programmed by other people, my robot self. *I* was not making the expensive phone call that would relieve me. The people who didn't "care" about us were the gossipers who criticize people for being "extravagant" and "neurotic." At this time, I had paid off my major debts incurred during illness, including fifteen months back rent. I had a thousand dollars in the bank. But I still was haunted by the years of being heavily in debt when I *was* criticized for expenditures and for being neurotic. The doctor I had not gone back to had told me that I was neurotic. Everyone knows that neurotic women plague doctors with telephone calls. I wasn't going to.

It was a completely unreal world that I was living in, because the previous doctor didn't think I was neurotic to call him when the cortisone needed to be changed, and as for the cost of the call, it would be less than the cost of going to the local doctor, which I would have felt no qualms about doing if I had had confidence in the doctor. When my mind gets mixed up, it just doesn't make sense.

When I understood my difficulty, I started toward the phone. But when I had taken a few steps, I stopped as though I had been stopped—not as though I had stopped myself. I couldn't go any farther. So I walked away, thought and did other things, and tried again. And got stopped again. When this happened, the effort of making myself take even one more step seemed too much, like trying to push a steam-roller out of the way. It seemed foolish to *try*. Then my mind said to me, over and over, that it was all right; that as long as I *understood* what was wrong and *knew* what had happened to me, everything was fine. It was very convincing, in spite of the fact that this told me nothing of what to do about the cortisone. This went on for two days before I fully knew that I *had* to put an end to it, that there *was* a battle going on in me that *I* had to win. I made myself go to the phone. When I heard the doctor's voice I said, with quavers that are not reproducible on paper, "This started out to be a medical call but it has wound up as a psychiatric one." Then we got around to the cortisone.

After that, I had some difficulty about calling the doctor, but none that I couldn't break through fairly easily. After the first several times, I phoned him if I needed his help, and not if I didn't, and that was all there was to it. This freedom was very beautiful to me, as it always is when I am free to act in accord with the total circumstances, on my *own* authority, not on what someone else thinks or says or has thought or has said that I or someone else should do. I live directly with the facts themselves.

In the beginning of my struggles, it was discouraging because I couldn't see the whole scene in the way that I express it now. I knew only that in *this* situation something was wrong and I had to correct it. Then there was another situation and I had to correct that. This went on . . . and on . . . and on . . . seemingly with nothing ahead and with no end to the going. But when I had gone through it enough times in different circumstances, then something that all the instances had in common began to show itself to me. I began to grasp in a total way the distinction between what others had put into me and what came out of myself. What had been a knotty tussle with one blindness after another, each one gone through in isolation from the others, began to be more flowing, with a more steady awareness of myself. Each time, something of myself came through, and something that was not myself got pushed away. There seems to be "no end to it" now, but the meaning of the words has changed. What began as one battle after another, so wearying, so full of pain, has now become frequently enjoyable, like the joy that a child has in his growing and in his growing knowing. Sometimes it isn't like that, but even then there is the knowing that *I* will come through, which certainly wasn't with me earlier, when I didn't even know what was pushing its way through. It is often true now that "I don't know *what* I'm gonna do, but *I'm* gonna do it"—not only in work (I don't know what I'll do when this book is done) and things like that, but in my relations with other people, too.

That client's words seem not to jibe with those of the one who said that he has to "Take more the position of passenger than driver," but to me it reads this way: When I make myself do what is not in accord with me, I am the driver, driving me—and often driving other people too. But the *real* driver comes from people outside me, telling me what to do. although I do not know that and think that I am doing it myself. I *feel* that I am the driver, but actually I am being driven. When my mind is cleared of outside intervention and I flow along with me, then I *feel* like a passenger, who doesn't have to clutch the wheel and watch the road. There is no car, no road, no driver.

When "growing opinion" says that I *am* the product of my class, this makes no sense to me. Certainly this can and does happen, but it doesn't *have* to happen. I have been conditioned, and there are ways in which I am still conditioned. But I can know when my conditioning is getting in my way, and I have been learning to do my own brain-washing. I have discovered that, although I can't yet do it all the time, still I *can*, for a short period—in some circumstances longer than in others—de-condition myself by flicking my mental switch and non-cold matter-of-factly deal with the facts and persons present. My "hope for education" includes the possibility that we might begin to learn to do this in the first grade, or even in kindergarten. It seems to me that if education were turned "upside down" it would find itself rightside up. Then, it would take place through the interaction of what is inside with what is outside, with the inside coming *first*. We seem to forget that that is where things came from in the first place.

When my daughter was twelve, I discovered that she had done a real and thorough job of research on the American cowboy. She had started out just by liking cowboy stories. She read every one that she could get hold of, quite indiscriminately, but then discrimination began to take place. The first one to go was Zane Grey because "I can tell at the beginning of the book how it's going to end." She became more interested in factual accounts, and in stories which were accurate in their information. In reading for her pleasure, she noticed that in different parts of the west, different names were used for the cowboys' gear and even for the cowboys themselves, who were "buckaroos" in Oregon. She noticed that the gear varied from place to place, according to the terrain and the influence of the Indians and the Spaniards. All this she had put down so neatly that it could be grasped a a glance, together with her own sketches where illustration was possible. For the first time, I thought of "research" as something not "out there" that must be learned from others, but something that in the first place came out of people in just the way that it had come out of her. How else could it have happened? A few years ago, I read a little book on the process of education, most of which was devoted to describing how learning takes place, based on studies in psychology. This was all very sound, as anyone could know by being with young children or by observing himself. But the rest of the book was devoted to the question, How can we *make* this happen? It seems to me that since that *is* what happens (learning), all that we have to do is *let* it happen. We're so back-end-to, it's pitiful. I have even heard teachers say that children have to be *taught* how to play.

An awful thing about what I am *taught* is that it doesn't grow. In school, I was taught that the past had changed—and for the better—but the present was "the end." We had arrived. So, the longer I lived, the more misinformation I had in my head—like 6% maximum legal interest, a country named Bohemia, the chief exports of Japan, and the English economic system—not to mention our own. What ours *is* still seems to me sometimes to be wrong *because* it isn't what I was taught it was. What I was *taught* it *was* is completely irrelevant.

Repeatedly, when I have discovered myself clinging tenaciously to something that isn't so, I have found that it is something that I was taught. What I learn myself is more flexible. My own observations may be fallible—I know they are—but they are a lot less fallible than anything I have been taught because *change* is a part of what I notice, and I don't get stuck.

An academic course has definite boundaries: even if I learn everything within them, I can't go beyond them, and when I have completed the *course* I feel that I *"know."* When I am studying on my own, there are no such boundaries. I range wherever my interests take me, into the disapproved as well as the approved, instead of being confined to texts and other reading selected for me by someone else. In this way, I gain a perspective that makes me a more reasonable person, more in accord with reality. I become aware of many divergent opinions and my mind is more flexible as I find my own continuing way through them. The continuing happens because this is not a course, with the end arbitrarily decided by someone else. There is no end. My awareness of this changes my behavior.

Relatedness (otherwise known as "a broad liberal background") comes about through one field moving me into another, not through certain things being put together and related *for* me. Within myself, this relatedness is more vast and sweeping than anything that can be put in books or taught to me by someone else. The space within me is prodigious. A person can look into himself and discover that—and also discover that this space shrinks when I have taken in too much without time to digest it. Then I become like the Navajos who say that "there's no more room in the head."

When I study on my own, I discover that while many things can be quite well known in a general way, they can't be accurately or permanently pinned down. I become less dogmatic, and at the same time more free, living with the uncertainty that is a reality of life. Through reading authors of many different periods, I notice how each has been conditioned by his times, and this leads me to seek out in which ways my own view is affected by my times. I notice where this is freeing and where it binds me, and then I can begin to cut the bonds, which are, I discover, not contemporary but a hangover from the past, prevalent but dying. I move, then, with what is truly contemporary, with what is appearing *now*—the living change, not the dying. I find authors whose views have changed in successive books, which tends to keep my mind more open about both of us. I discover that when I re-read a book out of my own interest, what it says to me the second time may be quite different from what it said the first time. This brings me closer to reality about myself and books. All this in itself has an effect on my interpersonal relations, apart from the fact that when I am ranging freely I am happy—not happy *about*, just happy—and that affects my interpersonal relations too.

These facts to me are *significant* learning. They are basic, universal, applicable to any people, place and time. When I am aware of them, I am in touch with the unchanging reality of change. With this awareness *first*, then what I do in the ephemeral world of my own lifetime is more intelligent, including my relations with other people. At the same time, I am a more autonomous person, able to find out for myself, and with trust in my ability to find my way. That my way includes the help of others in no way diminishes my independence because I do the choosing for myself. I accept what I can use at the time, what is meaningful *to me*. Then, all that I learn is linked together, inside *me*, with more connections than could be written down even if I spent a lifetime doing it, because new ones are constantly being made while I am writing. All of these connections are available to me through my inner computer, as I need them. One part of me is such a fantastic machine, contained in such a little space and so easy to take with me, that it is idiotic to get excited about the feats of machines that are made by men. They are convenient, if we use them properly, and that is all. I have to use my own machine properly too, by not interfering with it, because when it is interfered with it goes haywire.

It doesn't seem to me that a problem is *necessary* for this kind of learning, although a problem certainly can stimulate me. But perhaps I am using the word "problem" in a too limited way.

When I first learned to read, I was absorbed in the joy of my learning. At the same time, it is true that until I could read for myself, I was dependent on my mother's time: I had to wait for her to read to me. That could be called a problem. But when I was eight and went to German school on Saturday mornings, I was lost in the enchantment of writing German script, the light lines up, the heavy lines down. It was utter fascination to find that a completely different set of sounds and "pictures" (letters)—like *frau*—meant in another part of the world what my mother was. To discover that *kindergarten,* to which I had gone, was a German word, not an English one, was like discovering that I had been in Germany instead of America. Then there were the words that were the same as ours, and others that were *nearly* the same, which brought together what had seemed so far apart.

I do not think that I had a *problem* that led to my learning German. I did have some familiarity with it, to begin with. My German grandfather still spoke German when he had someone to speak it with. German was in this sense already a part of my life.

But what did my love of Latin in high school have to do with my life? The only connection that I can find is that I liked words, and the meanings of words, and the derivations of words. But after a year I had got the hang of that and could look up the rest for myself when I needed it.

How did my daughter become so interested in cowboys, when her life had been in Hawaii, New York City, England, and the New England coast—before the days of TV and when she hadn't seen many movies? I think the answer is that she loved *horses.* At three, when she said that she felt sick and I asked her what would make her well, she said, "A wide on de pony in de park!" When she was five, the school reported that she was doing well in piano and suggested that she have private lessons. I asked her if she would like this and she said, after only a moment's pause, "Well, I *would* rather ride a horse." She always wanted the horse she didn't have, and that may have been her problem, that she solved by being with horses in books.

When I went to live in Hawaii, for the first time I got interested in studying history. At that time, the written history was only about a hundred years old and all the sources were available to me. At the same time, I could question people for whom much of that past was a part of their own experience, and another lot of it had come to them through their immediate forebears. As I was living in the Islands, all this contributed to my understanding of what I was living *in,* and I liked this. I think another factor may have been that in the Islands I was exposed to a different way of thinking about life, and I got really interested in Island history in that sense.

If "problems" entered into this, I think it may have been chiefly that when I was first in Hawaii I *did* have problems, particularly during the first six months. Although I had driven in New York Labor Day traffic and all that, when I got to Hawaii the traffic in Honolulu scared me so that I parked on Punchbowl and walked into town, until I could figure out how to drive in it. Time after time, as I went around in Honolulu doing errands, I watched the traffic to try to get the hang of it. At last it became clear to me: instead of going by *rules,* the drivers went by *noticing.* They noticed other cars and they noticed pedestrians and moved in accord with them. Then I drove into town and drove as they did and enjoyed it. I had a problem with the *slowness* of everything in Hawaii: it was what I wanted for myself, but it was still a problem to me because I got irked by it and had to find my way out of that. I had to slow down *myself: then* I enjoyed the ease and lack of friction in moving slowly. I had the problem of living where it seemed to me that *everyone* had dark hair, dark skin, dark eyes. It wasn't race prejudice, just homesickness. When I returned to the white skins on the mainland eight years later, it seemed to me that everyone looked sick. I loved the mangoes, papayas, soursops, guavas, mountain apples, cherimoyas, pohas, and all the rest, but it seemed to me that if I could get to California (a place I'd never really liked) and eat apricots, peaches, grapes, pears, apples for ten days, then it would be easier for me to go on living in the Islands. I had the problem of ants: I thought I must be a sloppy housekeeper because there were always ants in my house. Sometimes I dived in to help people—and got bumped. The people were somehow different from those I had lived with before.

There were all these re-evaluations and adjustments going on, and this seems to open up my mind and lead to questions in other areas too. In trying to answer them, I *run into* problems, but it doesn't seem *necessary* for me to have the problem first. This suggests to me that in trying to provide our children with "security" by keeping them within a close environment we may be stupidizing them. Then, when we make them specialize at an early age, we are stupidizing them still further.

But is it *only* the "problem," anyway? It seems to me that I could have resolved my problem of being sick in another way, but when I found out how little medicine knows about the whole field of chronic illness, I was intrigued by the unknown territory that I lived in. It scared me at first, but then I began to get used to it, and I *like* moving in the unknown, even when it scares me and is painful. I don't like the scare and the pain, but I *do* like the exploring. With the doctor, I explored what is known by medicine about chronic illness, with particular reference to my own. Through his honesty, I learned how *much* is *not* known. All this churned around in my head until I had the feeling of a dead end, of being somehow on the wrong track. That was when I sideslipped into psychology. It happened. This could lead me back into medicine sometime. That happens too.

This suggests that telling children what is *not* known may be as helpful as telling them what *is*. When I am told only what is known, it seems to be "all there"—nothing that *I* can contribute to it. I can only learn what others have found out. It is a closed world. When I know how small the known is in relation to the unknown, the whole world opens up before me. I am free to explore and make my own discoveries instead of being a passive recipient of what is known—to the point that I think that *everything* is known by someone else and that it is only through others that I can acquire knowledge.

I have seen this happen recently with several people, individually, when they discovered how chaotic the field of psychotherapy is, how much is not known. One of them said, with such freedom, "Each man is on his own!" and the others expressed themselves in much the same way. They moved from the limitation of trying to find a specific answer already known to the freedom of making their own discoveries. More is known about how we get into the troubles that we do than is known about how to get out of them. With this knowledge of how things go wrong, these people are now trying to find out for themselves how to go about the undoing, both in the sense of curing and of prevention, by intelligent trial and error. If we all did this, it seems to me that we would get out of our difficulties a lot faster. Some people are very afraid of this, because people will make mistakes.

That certainly will happen. But these will not be such *persistent* mistakes as *taught* mistakes are, and I think it is the *persistent* mistakes that foul us up. It seems to me that much damage has been done by psychotherapists who believed that because they had been trained they knew the answers, and so went on and on with their mistakes. Being sure that the answers they had were right, they didn't ask enough questions, and when things didn't go well, the questions that arose were answered in terms of the answers they already had. That's a good way to get stuck. If *all* psychotherapists had done that, we wouldn't have arrived at the new concepts that we have today.

Holding to a belief seems to be a large part of our trouble. If I am sure that beating a child is good, or that permissiveness is good, poor results will only lead me to push harder with whatever it is that I believe. That seems to have happened with education. "It hasn't produced what we wanted, so let's have more of it" seems to sum up what I hear from all kinds of people all the time.

When a profession makes a mistake, a whole generation or more is damaged by it—like four-hour feedings and no rocking the baby. An individual's mistake just doesn't reach that far.

Another good thing about everyone trying to work out psychotherapy on his own (and including himself) is that all these people wouldn't be Authorities. If I try out something in my relations with other people and it doesn't work well for them, they will be resistant to me in a way they would not be to an authority, who is believed to know what he is doing.

It is my experience of most authorities that "working with them" means that I do as they say even if it gives me hell. That's just slavery in another form. "Working with authorities" makes sense to me only when the professional and the amateur each say, "Of what you offer, I go along with this, but not with that," or "Let's try this and see how it works out." Then, there is continuing education on both sides, and each one at the same time is responsible for himself, through the choices he makes. That actually makes it much easier on the authorities, and it seems to me that any authority in his right mind would welcome this: he has contributed the best that he has to offer, but the *choice* is made by the other person. The authority, likewise, is responsible for *his* choices —for himself, not for anyone else. I know how this works out because I've tried it in medicine, education, and psychology. When I meet with authorities who permit this, I become more intelligent and I enjoy democracy. When I, a parent-authority, permit this, I become more intelligent and I enjoy democracy. Children, in my experience, become more responsible when they are made clearly responsible for themselves. When they see the responsibility as being up to someone else, they are less so. This happens to grownups, too.

It seems to me that democracy hasn't failed: we just haven't really tried it yet. As a boy in grade school said to his teacher, "Mr. T—, how do you reconcile your teaching of democracy with the way that you conduct this class?"

We need to be more congruent.

IV

I knew the mass of men concealed
Their thoughts, for fear that if revealed
They would by other men be met
With blank indifference, or with blame reproved;
I knew they lived and moved
Tricked in disguises, alien to the rest
Of men, and alien to themselves. . . .

Matthew Arnold *"The Buried Life"*

It's a crazy gift we have, this trickery. My inside knowing of it is remembered only as far back as my own third year when:

My mother and father laugh at me because I am enchanted by a hole in the ground. The hole is being dug on the next street so I cannot go there alone. I wait with excitement for my father to take me. When he does, I look into that ever-deepening hole with the same fascination that I watch my mother peel potatoes, noticing the changing form, the changing color, the changing texture, and the changing fragrance of both holes in the ground and potatoes.

My father tells my mother, "A hole in the ground!" (The way that he says it, I know this is not much.) "You'd think there was a magnet at the bottom. If I didn't hold her hand, she'd tumble right in." (I don't catch all those words at the time. There are too many that I don't know. I hear them later on, when my father tells someone, and I remember my pain, and what I did about it.)

My mother and father laugh together and are tender with me and love me, but they do not understand. I feel alone, and my enchantment is bleeding around the edges. I am the angry which is hurt. Being at the moment true to me, I scowl at my parents.

They jolly me then, because children must be kept happy. And then I am not true to me. I laugh, because in that grown-up world of which I would like to be a part, that is the thing to do. (A few years later, when I am attracted by a hole in the ground, I drop a marble into it, so that if anyone comes along I can say that I am looking for my marble, not that I am enjoying the hole, which would be ridiculed.)

There are times when I scowl at my parents not because I am misunderstood, or not understood, but because I have discovered that that is a way to get their attention. And then, when they have brought me around from scowls to laughter they are very pleased with both themselves and me.

And I am pleased with myself for having figured this out.

That's a long way from being happy with a hole in the ground.

Less than three years in this world, and I have got involved in cleverness. I didn't develop that all by myself. Already my parents have been tricking *me,* and I have watched other trickery go on with aunts and uncles, grandpa. . . .

My parents and I love each other and enjoy each other. Most of the time we are sensitive to each other at some level. We don't know that increasingly we are being superficial, that there is a dimension missing, and that their lack of respect for me is developing in me a lack of respect for them. They respect me in the outside things, like letting me paint the railings on the porch and carry things that would break if I dropped them, but they don't respect my insides because they think I haven't any.

When I am playing on the floor, they talk together or with other people about things I am "too little to understand" and so it is all right to say them. But they don't know what my understanding is, and I have no way to tell them. So it goes round and round inside my head. Sometimes I understand some things, and I am hurt. Other times, I do not understand and I try to put things together, with not enough information in my head to arrive at making sense. So what I make of it is nonsense, but I don't know that. And sometimes what I make of it won't stay put. I think of it one way and I hurt, so I think of it another way so I'll feel comfortable. But it slips around, and I don't know which way it *is*. I am too young to know that the word for my trouble is confusion.

To my parents, aunts and uncles, my life is good: I have loving parents, aunts and uncles, and "How blessed it is to be a child and have no worries."

To me, it often seems that I must have been born to the wrong parents, or that these are not my parents, because my *real* parents would know *me*.

My sister, six years older than I, bewilders me, because sometimes she is a child with me; and then suddenly she switches and talks like a grownup. She tells me what to think and feel. What *I* think and feel she says is silly. A moment ago, she was agreeing with me. Sometimes I fight my sister about this. But sometimes I say that I think and feel what she does, and then I feel BIG.

But then I get all mixed up and I am crying "Who am I?"

And no one helps me with that because it is a silly question. I'm me. Who else could I be? This seems so to me, too, so why is it that I don't know? There must be something so wrong with me that no one will tell me about it.

I talk to my puppy and my dolls, and to the trees. They don't confuse me because they listen, and I can say *anything* I want and they don't talk back. They go on listening. And then I begin to hear myself and know it's me.

*The Interpersonal Relationship: The Core of Guidance**

Carl R. Rogers

I would like to share with you in this paper a conclusion, a conviction, which has grown out of years of experience in dealing with individuals, a conclusion which finds some confirmation in a steadily growing body of empirical evidence. It is simply that in a wide variety of professional work involving relationships with people—whether as a psychotherapist, teacher, religious worker, guidance counselor, social worker, clinical psychologist—it is the *quality* of the interpersonal encounter with the client which is the most significant element in determining effectiveness.

Let me spell out a little more fully the basis of this statement in my personal experience. I have been primarily a counselor and psychotherapist. In the course of my professional life I have worked with troubled college students, with adults in difficulty, with "normal" individuals such as business executives, and more recently with hospitalized psychotic persons. I have endeavored to make use of the learnings from my therapeutic experience in my interactions with classes and seminars, in the training of teachers, in the administration of staff groups, in the clinical supervision of psychologists, psychiatrists, and guidance workers as they work with their clients or patients. Some of these relationships are long-continued and intensive, as in individual psychotherapy. Some are brief, as in experiences with workshop participants or in contacts with students who come for practical advice. They cover a wide range of depth. Gradually I have come to the conclusion that one learning which applies to all of

* *Harvard Educational Review:* Vol. 32, No. 4, Fall 1962.

these experiences is that it is the quality of the personal relationship which matters most. With some of these individuals I am in touch only briefly, with others I have the opportunity of knowing them intimately, but in either case the quality of the personal encounter is probably, in the long run, the element which determines the extent to which this is an experience which releases or promotes development and growth. I believe the quality of my encounter is more important in the long run than is my scholarly knowledge, my professional training, my counseling orientation, the techniques I use in the interview. In keeping with this line of thought, I suspect that for a guidance worker also the relationship he forms with each student—brief or continuing—is more important than his knowledge of tests and measurements, the adequacy of his record keeping, the theories he holds, the accuracy with which he is able to predict academic success, or the school in which he received his training.

In recent years I have thought a great deal about this issue. I have tried to observe counselors and therapists whose orientations are very different from mine, in order to understand the basis of their effectiveness as well as my own. I have listened to recorded interviews from many different sources. Gradually I have developed some theoretical formulations, some hypotheses as to the basis of effectiveness in relationships. As I have asked myself how individuals sharply different in personality, orientation and procedure can all be effective in a helping relationship, can each be successful in facilitating constructive change or development, I have concluded that it is because they bring to the helping relationship certain attitudinal ingredients. It is these that I hypothesize as making for effectiveness, whether we are speaking of a guidance counselor, a clinical psychologist, or a psychiatrist.

What are these attitudinal or experiential elements in the counselor which make a relationship a growth-promoting climate? I would like to describe them as carefully and accurately as I can, though I am well aware that words rarely capture or communicate the qualities of a personal encounter.

Congruence

In the first place, I hypothesize that personal growth is facilitated when the counselor is what he *is,* when in the relationship with his client he is genuine and without "front" or facade, openly being the feelings and attitudes which at that moment are flowing in him. We have used the term "congruence" to try to describe this condition. By this we mean that the feelings the counselor is experiencing are available to him, available to his awareness, that he is able to live these feelings, be them in the relationship, and able to communicate them if appropriate. It means that he comes into a direct personal encounter with his client, meeting him on a person-to-person basis. It means that he is *being* himself, not denying himself. No one fully achieves this condition, yet the more the therapist is able to listen acceptantly to what is going on within himself, and the more he is able to *be* the complexity of his feelings without fear, the higher the degree of his congruence.

I think that we readily sense this quality in our everyday life. We could each of us name persons whom we know who always seem to be operating from behind a front, who are playing a role, who tend to say things they do not feel. They are exhibiting incongruence. We do not reveal ourselves too deeply to such people. On the other hand each of us knows individuals whom we somehow trust, because we sense that they are being what they *are*, that we are dealing with the person himself, and not with a polite or professional facade. This is the quality of which we are speaking, and it is hypothesized that the more genuine and congruent the therapist is in the relationship, the more probability there is that change in personality in the client will occur.

I have received much clinical confirmation for this hypothesis in recent years in our work with randomly selected hospitalized schizophrenic patients. The individual therapists in our research program who seem to be most successful in dealing with these unmotivated, poorly educated, resistant, chronically hospitalized individuals, are those who are first of all real, who react in a genuine, human way as persons, and who exhibit their genuineness in the relationship.

But is it always helpful to be genuine? What about negative feelings? What about the times when the counselor's real feeling toward his client is one of annoyance, or boredom, or dislike? My tentative answer is that even with such feelings as these, which we all have from time to time, it is preferable for the counselor to be real than to put up a facade of interest and concern and liking which he does not feel.

But it is not a simple thing to achieve such reality. I am not saying that it is helpful to blurt out impulsively every passing feeling and accusation under the comfortable impression that one is being genuine. Being real involves the difficult task of being acquainted with the flow of experiencing going on within oneself, a flow marked especially by complexity and continuous change. So if I sense that I am feeling bored by my contacts with this student, and this feeling persists, I think I owe it to him and to our relationship to share this feeling with him. But here again I will want to be constantly in touch with what is going on in me. If I am, I will recognize that it is *my* feeling of being bored which I am expressing, and not some supposed fact about him as a boring person. If I voice it as *my own* reaction, it has the potentiality of leading to a deeper relationship. But this feeling exists in the context of a complex and changing flow, and this needs to be communicated too. I would like to share with him my distress at feeling bored, and the discomfort I feel in expressing this aspect of me. As I share these attitudes I find that my feeling of boredom arises from my sense of remoteness from him, and that I would like to be more in touch with him. And even as I try to express these feelings, they change. I am certainly *not* bored as I try to communicate myself to him in this way, and I am far from bored as I wait with eagerness and perhaps a bit of apprehension for his response. I also feel a new sensitivity to him, now that I have shared this feeling which has been a barrier between us. So I am very much more able to hear the surprise or perhaps the hurt in his voice as he now finds *him*self speaking more genuinely because I have dared to be real with him. I have let myself be a person—real, imperfect—in my relationship with him.

I have tried to describe this first element at some length because I regard it as highly important, perhaps the most crucial of the conditions I will describe, and because it is neither easy to grasp nor to achieve. Gendlin has done an excellent job of explaining the significance of the concept of experiencing and its relationship to counseling and therapy, and his paper may supplement what I have tried to say.

I hope it is clear that I am talking about a realness in the counselor which is deep and true, not superficial. I have sometimes thought that the word transparency helps to describe this element of personal congruence. If everything going on in me which is relevant to the relationship can be seen by my client, if he can see "clear through me," and if I am *willing* for this realness to show through in the relationship, then I can be almost certain that this will be a meaningful encounter in which we both learn and develop.

I have sometimes wondered if this is the only quality which matters in a counseling relationship. The evidence seems to show that other qualities also make a profound difference and are perhaps easier to achieve. So I am going to describe these others. But I would stress that if, in a given moment of relationship, they are not genuinely a part of the experience of the counselor, then it is, I believe, better to be genuinely what one is, than to pretend to be feeling these other qualities.

Empathy

The second essential condition in the relationship, as I see it, is that the counselor is experiencing an accurate empathic understanding of his client's private world, and is able to communicate some of the significant fragments of that understanding. To sense the client's inner world of private personal meanings as if it were your own, but without ever losing the "as if" quality, this is empathy, and this seems essential to a growth-promoting relationship. To sense his confusion or his timidity or his anger or his feeling of being treated unfairly as if it were your own, yet without your own uncertainty or fear or anger or suspicion getting bound up in it, this is the condition I am endeavoring to describe. When the client's world is clear to the counselor and he can move about in it freely, then he can both communicate his understanding of what is vaguely known to the client, and he can also voice meanings in the client's experience of which the client is scarcely aware. It is this kind of highly sensitive empathy which seems important in making it possible for a person to get close to himself and to learn, to change and develop.

I suspect that each of us has discovered that this kind of understanding is extremely rare. We neither receive it nor offer it with any great frequency. Instead we offer another type of understanding which is very different, such as "I understand what is wrong with you" or "I understand what makes you act that way." These are the types of understanding which we usually offer and receive—an evaluative understanding from the outside. It is not surprising that we shy away from true understanding. If I am truly open to the way life is experienced by another person—if I can take his world into mine—then I run the risk of seeing life in his way, of being changed myself, and we all resist change. So we tend to view this other person's world only in our terms, not in his. We analyze and evaluate it. We do not understand it. But when someone understands how it feels and seems to be me, without wanting to analyze me or judge me, then I can blossom and grow in that climate. I am sure I am not alone in that feeling. I believe that when the counselor can grasp the moment-to-moment experiencing occurring in the inner world of the client, as the client sees and feels it, without losing the separateness of his own identity in this empathic process, then change is likely to occur.

Though the accuracy of such understanding is highly important, the communication of intent to understand is also helpful. Even in dealing with the confused or inarticulate or bizarre individual, if he perceives that I am *trying* to understand his meanings, this is helpful. It communicates the value I place on him as an individual. It gets across the fact that I perceive his feelings and meanings as being *worth* understanding.

None of us steadily achieves such a complete empathy as I have been trying to describe, any more than we achieve complete congruence, but there is no doubt that individuals can develop along this line. Suitable training experiences have been utilized in the training of counselors, and also in the "sensitivity training" of industrial management personnel. Such experiences enable the person to listen more sensitively, to receive more of the subtle meanings the other person is expressing in words, gesture, and posture, to resonate more deeply and freely within himself to the significance of those expressions.*

* I hope the above account of an empathic attitude will make it abundantly clear that I am not advocating a wooden technique of pseudo-understanding in which the counselor "reflects back what the

Positive Regard

Now the third condition. I hypothesize that growth and change are more likely to occur the more that the counselor is experiencing a warm, positive, acceptant attitude toward what *is* in the client. It means that he prizes his client, as a person, with somewhat the same quality of feeling that a parent feels for his child, prizing him as a person regardless of his particular behavior at the moment. It means that he cares for his client in a non-possessive way, as a person with potentialities. It involves an open willingness for the client to be whatever feelings are real in him at the moment—hostility or tenderness, rebellion or submissiveness, assurance or self-depreciation. It means a kind of love for the client as he is, providing we understand the word love as equivalent to the theologian's term *agape,* and not in its usual romantic and possessive meanings. What I am describing is a feeling which is not paternalistic, nor sentimental, nor superficially social and agreeable. It respects the other person as a separate individual, and does not possess him. It is a kind of liking which has strength, and which is not demanding. We have termed it positive regard.

Unconditionality of Regard

There is one aspect of this attitude of which I am somewhat less sure. I advance tentatively the hypothesis that the relationship will be more effective the more the positive regard is unconditional. By this I mean that the counselor prizes the client in a total, rather than a conditional way. He does not accept certain feelings in the client and disapprove others. He feels an *unconditional* positive regard for this person. This is an outgoing, positive feeling without reservations and without evaluations. It means *not* making judgments. I believe that when this nonevaluative prizing is present in the encounter between the counselor and his client, constructive change and development in the client is more likely to occur.

Certainly one does not need to be a professional to ex-

client has just said." I have been more than a little horrified at the interpretation of my approach which has sometimes crept into the teaching and training of counselors.

perience this attitude. The best of parents show this in abundance, while others do not. A friend of mine, a therapist in private practice on the east coast, illustrates this very well in a letter in which he tells me what he is learning about parents. He says:

> I am beginning to feel that the key to the human being is the attitudes with which the parents have regarded him. If the child was lucky enough to have parents who have felt proud of him, wanted him, wanted him just as he was, exactly as he was, this child grows into adulthood with self-confidence, self-esteem; he goes forth in life feeling sure of himself, strong, able to lick what confronts him. Franklin Delano Roosevelt is an example . . . "my friends. . . ." He couldn't imagine anyone thinking otherwise. He had two adoring parents. He was like the pampered dog who runs up at you, frisking his tail, eager to love you, for this dog has never known rejection or harshness. Even if you should kick him, he'll come right back at you, his tail friskier than ever, thinking you're playing a game with him and wanting more. This animal cannot imagine anyone disapproving or disliking him. Just as unconditional regard and love was poured into him, he has it now to give out. If a child is lucky enough to grow up in this unconditionally accepting atmosphere, he emerges as strong and sure and he can approach life and its vicissitudes with courage and confidence, with zest and joy of expectation.
>
> But the parents who like their children—if. They would like them if they were changed, altered, different; if they were smarter or if they were better, or if, if, if. The offspring of these parents have trouble because they never had the feeling of acceptance. These parents don't really like these children; they would like them if they were like someone else. When you come down to the basic fundamental, the parent feels: "I don't like *this* child, this child before me." They don't say that. I am beginning to believe that it would be better for all concerned if parents did. It wouldn't leave such horrible ravages on these unaccepted children. It's never done that crudely.

> "If you were a nice boy and did this, that and the other thing, then we would all love you." I am coming to believe that children brought up by parents who would like them "if" are never quite right. They grow up assuming that their parents are right and that they are wrong; that somehow or other they are at fault; and even worse, very frequently they feel they are stupid, inadequate, inferior.

This is an excellent contrast between an unconditional positive regard and a conditional regard. I believe it holds as true for counselors as for parents.

The Client's Perception

Thus far all my hypotheses regarding the possibility of constructive growth have rested upon the experiencing of these elements by the counselor. There is, however, one condition which must exist in the client. Unless the attitudes I have been describing have been to some degree communicated to the client, and perceived by him, they do not exist in his perceptual world and thus cannot be effective. Consequently it is necessary to add one more condition to the equation which I have been building up regarding personal growth through counseling. It is that when the client perceives, to a minimal degree, the genuineness of the counselor and the acceptance and empathy which the counselor experiences for him, then development in personality and change in behavior are predicted.

This has implications for me as a counselor. I need to be sensitive not only to what is going on in me, and sensitive to the flow of feelings in my client. I must also be sensitive to the way he is receiving my communications. I have learned, especially in working with more disturbed persons, that empathy can be perceived as lack of involvement; that an unconditional regard on my part can be perceived as indifference; that warmth can be perceived as a threatening closeness, that real feelings of mine can be perceived as false. I would like to behave in ways, and communicate in ways which have clarity for this specific person, so that what I am experiencing in relationship to him would be perceived unambiguously by him. Like the other conditions I have proposed, the principle is easy to grasp; the achievement of it is difficult and complex.

Some Limitations

I would like to stress that these are hypotheses. In a later section I will comment on the way these hypotheses are faring when put to empirical test. But they are beginning hypotheses, not the final word.

I regard it as entirely possible that there are other conditions which I have not described, which are also essential. Recently I had occasion to listen to some recorded interviews by a young counselor of elementary school children. She was very warm and positive in her attitude toward her clients, yet she was definitely ineffective. She seemed to be responding warmly only to the superficial aspects of each child and so the contacts were chatty, social and friendly, but it was clear she was not reaching the real person of the child. Yet in a number of ways she rated reasonably high on each of the conditions I have described. So perhaps there are still elements missing which I have not captured in my formulation.

I am also aware of the possibility that different kinds of helping relationships may be effective with different kinds of people. Some of our therapists working with schizophrenics are effective when they appear to be highly conditional, when they do *not* accept some of the bizarre behavior of the psychotic. This can be interpreted in two ways. Perhaps a conditional set is more helpful with these individuals. Or perhaps—and this seems to me to fit the facts better—these psychotic individuals perceive a conditional attitude as meaning that the therapist *really* cares, where an unconditional attitude may be interpreted as apathetic noncaring. In any event, I do want to make it clear that what I have given are beginning formulations which surely will be modified and corrected from further learnings.

The Philosophy which is Implicit

It is evident that the kinds of attitudes I have described are not likely to be experienced by a counselor unless he holds a philosophy regarding people in which such attitudes are congenial. The attitudes pictured make no sense except in a context of great respect for the person and his potentialities. Unless the primary element in the counselor's value system is the worth of the individual, he is not apt to find himself experiencing a real caring, or a desire to understand, and perhaps he will not respect himself enough to be real. Certainly the professional person who holds the view that individuals are essentially objects to be manipulated for the welfare of the state, or the good of the educational institution, or "for their own good," or to satisfy his own need for power and control, would not experience the attitudinal elements I have described as constituting growth-promoting relationships. So these conditions are congenial and natural in certain philosophical contexts but not in others.

Empirical Studies

This raises some questions which I have asked myself, and which you too must be asking. Are these characteristics which I have described as essential to a helping relationship simply my personal opinion, preference, and bias? Or do they represent simply a bias growing out of a generally democratic philosophy? Or do they in *fact* promote constructive change and development?

Five years ago I could not have answered these questions. Now there are at least a dozen well-designed research investigations which, approaching the matter in a variety of ways, throw light on the issues. To report each of these studies would be confusing rather than helpful. Let me try to describe their methods in general terms and then report on the findings.

The studies deal with two rather different classes of clients: students and community members who voluntarily come to counselors for help; and on the other hand, schizophrenic individuals in a state hospital who have been there for periods ranging from a few months to many years. The first group is above the socio-educational average, the second below. The first group is motivated to gain help, the second is not only unmotivated but resistant. The over-all range in adjustment is from well-functioning individuals through varying degrees of maladjustment and disturbance, to those who are completely unable to cope with life, and who are out of contact with reality.

In the different studies there have been three ways of measuring the attitudinal elements I have described. The first method is based on brief segments, usually four minutes in length, taken in a randomized way from the tape-recorded interviews. Raters, listening to these segments, judge the degree to which the counselor is, for example, being accurately empathic, and make a rating on a carefully defined scale. The raters have no knowledge of whether the segment is from an early or late interview, or whether it is a more or less successful case. In most of the studies several raters have made ratings on each of the qualities involved.

A second method of measurement is through the use of the Relationship Inventory, filled out by the client at different points in time. The Inventory contains statements regarding the degree to which the counselor is acceptant, empathic, and congruent, and the client responds by evaluating the statement on a six point scale from "strongly true" to "definitely untrue." Examples concerning empathy are: "He generally senses or realizes how I am feeling;" "He understands my words but does not realize how I feel." In relationship to congruence some items are: "He behaves just the way that he is in our relationship;" "He pretends that he likes me or understands me more than he really does." The Inventory is scored for each of the four attitudinal elements, and there is also a total score.

The third method is also based on the Relationship Inventory, but this time filled out by the therapist or counselor. The items are identical except for a suitable change in pronouns.

In the various studies, different criteria are used for assessing the degree of constructive personality change which has taken place over the course of the interviews. In all cases the criteria of change are independent of the measurement of attitudinal conditions in the relationship. Some of the measures of change are: changes in various Minnesota Multiphasic Personality Inventory scales and indices; changes in projective tests as analyzed "blind" by clinicians having no knowledge of the research; changes in Q-sort adjustment score; changes on a measure of anxiety; therapist's ratings of change in personality and in adjustment.

The Findings

Let me now give some of the general findings from these studies:

The counselor is the most significant factor in setting the level of conditions in the relationship, though the client, too, has some influence on the quality of the relationship.

Clients who will later show more change perceive more of these attitudinal conditions early in the relationship with their counselor or therapist.

The more disturbed the client, the less he is likely to (or able to?) perceive these attitudes in the counselor.

Counselors or therapists tend to be quite consistent in the level of attitudinal conditions which they offer to each client.

The major finding from all of the studies is that those clients in relationships marked by a high level of counselor congruence, empathy and unconditional positive regard, show constructive personality change and development. These high levels of conditions are associated with: positive change on MMPI scales and indices, including ego-strength; positive change from the pre- to post-test battery as rated by clinicians working "blind"; decrease in anxiety scores and in a self-consciousness score; a higher level on Process Scales designed to measure process in therapy; and positive change in counselor's ratings.

Clients in relationships characterized by a low level of these attitudinal conditions show significantly less positive change on these same indices.

In studies of clinic clients the correlation between the client's perception of the conditions offered early in the relationship and the degree of change at the conclusion of the interviews is somewhat higher than that between the counselor's perception of the conditions offered and the degree of change. The client's perception is, in other words, the better predictor of change.

This finding does not hold for the schizophrenic client, whose inner disturbance makes it difficult for him accurately to perceive the conditions offered by our conscientious and experienced therapists. With our schizophrenics, the rating of the conditions made by unbiased raters is the best predictor of change.

An unexpected finding with the schizophrenic clients is

that low conditions in the relationship are associated with *negative* change in several respects. The clients not only fail to show constructive change but become worse in the judgment of clinicians rating their pre- and post-test batteries; show an increase in anxiety; are worse off than their matched no-therapy controls. Whether this finding holds for clinic clients who come for help has not yet been determined.

A finding which seems to lend validity to the studies is that, as might be expected, more experienced counselors, when compared with inexperienced counselors, offer a higher level of these conditions, and are more successful in communicating these to their clients. Thus they are perceived as offering higher conditions, and their clients show more change over the course of the interviews.

Implications

What are some of the implications of these hypotheses and of these findings for the field of counseling psychology and guidance? I would like to mention four which occur to me.

In the first place, these studies indicate that perhaps it is possible to study cause and effect in counseling and psychotherapy. They are actually, so far as I know, the first studies to endeavor to isolate and measure the primary change-producing influences in counseling. Whether they are still further confirmed by later research, or whether they are contradicted or modified by future studies, they represent pioneering investigations of the question, "What really makes the difference in counseling and psychotherapy?" And the answer they give is that it is the attitudes provided by the counselor, the psychological climate largely created by him, which *really* makes the difference, which really induces change.

There is another highly practical significance to these studies. They indicate quite clearly that, by assessing a relationship early in its existence, we can to some degree predict the probability of its being a relationship which makes for growth. It seems to be quite within the range of possibility that in the not too distant future we will acquire an increasingly accurate knowledge of the elements which make for constructive psychological development, just as we have in the realm of nutrition acquired an increasingly accurate knowledge of the elements which promote physical growth. As this knowledge accumulates, and as our instruments grow sharper, then there is the exciting possibility that we may be able, relatively early in the game, to predict whether a given relationship will actually promote or inhibit individual psychological growth and development, just as we can assess the diet of a child and predict the extent to which this diet will promote or inhibit physical growth.

In this connection the disturbing finding that an inadequate interpersonal relationship can have a negative effect on personal development, at least in the case of highly disturbed individuals, makes such early assessment of a relationship an even more challenging possibility and responsibility.

Another significant meaning for the counseling field is that we now have the beginnings of a theory, and some empirical facts supporting the theory, as to the specific elements in an interpersonal relationship which facilitate positive change. Thus we can now say with some assurance and factual backing that a relationship characterized by a high degree of congruence or genuineness in the counselor, by sensitive and accurate empathy on the part of the counselor, by a high degree of regard, respect and liking for the client by the counselor, and by an absence of conditionality in this regard, will have a high probability of being an effective, growth-promoting relationship. This statement holds, whether we are speaking of maladjusted individuals who come of their own initiative seeking help, or whether we are speaking of chronically schizophrenic persons with no conscious desire for help. This statement also holds whether these attitudinal elements are rated by impartial observers who listen to samples of the recorded interviews, or whether they are measured in terms of the counselor's perception of the qualities he has offered in the relationship or whether they are measured by the client's perception of the relationship, at least in the case of the nonhospitalized client. To me it seems to be quite a forward stride to be able to make statements such as these in an area as complex and subtle as the field of helping relationships.

Finally, these studies would, if confirmed by further work, have significant implications for the training of counselors and therapists. To the extent that the counselor is seen as being involved in interpersonal relationships, and to the extent that the goal of those relationships is to promote healthy development, then certain conclusions would seem to follow. It would mean that we would endeavor to select individuals for such training who already possess, in their ordinary relationships with other people, a high degree of the qualities I have described. We would want people who were warm, spontaneous, real, understanding, and non-judgmental. We would also endeavor so to plan the educational program for these individuals that they would come increasingly to *experience* empathy and liking for others, and that they would find it increasingly easier to be themselves, to be real. By feeling understood and accepted in their training experiences, by being in contact with genuineness and absence of facade in their instructors, they would grow into more and more competent counselors. There would be as much focus in such training on the interpersonal experience as on the intellectual learning. It would be recognized that no amount of knowledge of tests and measures, or of counseling theories, or of diagnostic procedures could make the trainee more effective in his personal encounter with his clients. There would be a heavy stress upon the actual experience of working with clients, and the thoughtful and self-critical assessment of the relationships formed.

When I ask myself whether the training programs which I know, in guidance, in clinical psychology, in psychiatry, approach this goal, I come up with a strong negative. It seems to me that most of our professional training programs make it *more* difficult for the individual to be himself, and more likely that he will play a professional role. Often he becomes so burdened with theoretical and diagnostic baggage that he becomes *less* able to understand the inner world of another person as it seems to that person. Also, as his professional training continues, it all too often occurs that his initial warm liking for other persons is submerged in a sea of diagnostic and psychodynamic evaluation.

Thus to take the findings of these studies seriously would mean some sharp changes in the very nature of professional training, as well as in its curriculum.

Conclusion

Let me conclude with a series of statements which for me follow logically one upon the other.

The purpose of most of the helping professions, including guidance counseling, is to enhance the personal development, the psychological growth toward a socialized maturity, of its clients.

The effectiveness of any member of the profession is most adequately measured in terms of the degree to which, in his work with his clients, he achieves this goal.

Our knowledge of the elements which bring about constructive change in personal growth is in its infant stages.

Such factual knowledge as we currently possess indicates that a primary change-producing influence is the degree to which the client experiences certain qualities in his relationship with his counselor.

In a variety of clients—normal, maladjusted, and psychotic—with many different counselors and therapists, and studying the relationships from the vantage point of the client, the therapist, or the uninvolved observer, certain qualities in the relationship are quite uniformly found to be associated with personal growth and change.

These elements are not constituted of technical knowledge or ideological sophistication. They are personal human qualities—something the counselor *experiences*, not something he *knows*. Constructive personal growth is associated with the counselor's realness, with his genuine and unconditional liking for his client, with his sensitive understanding of his client's private world, and with his ability to communicate these qualities in himself to his client.

These findings have some far-reaching implications for the theory and practice of guidance counseling and psychotherapy, and for the training of workers in these fields.

V

I have discovered that each mother is on her own, with *each* child.

Betty Michael

Each time I study the preceding paper in relation to myself, I am more deeply in accord with it—except for tripping over the first paragraph in the letter from Carl Rogers' eastern therapist friend. I don't believe it. The Spanish *no me acuerdo* is a better way to put it: "This is not in accord with me"—my own knowing. Carl may not go along with it either by now, and the eastern therapist himself may have changed, so I am not arguing with *them* but with these words on paper.

Some of my reaction is prejudice. I thought that Roosevelt did some very fine things but that he tended to confuse himself with God. When a little boy came to my door and said:

Mrs. Stevens, the President is dead.

Ohh? I said. What did he die of?

I think there was something wrong with his brain, the small boy said.

Douglas, I said, I've been thinking that for some time.

In view of my prejudice, any discussion of this paragraph in terms of Roosevelt is meaningless. It would be anyway, because I never knew him, and in particular I never knew him closely, and I don't really know anyone unless we have been close. Their outsides, yes. Their insides, no.

Leaving Roosevelt out of it, this paragraph does not match with me except on terms which I don't accept. The friendly puppy people whom I have known have lived to my age and beyond without having to go to a psychotherapist, without cracking up, without being an exceptional nuisance to their friends and neighbors, and they have made a *lot* of sense (in my view). When a father was dressing down his son because he was half an hour late returning from a party, the mother said to me, "I'm just so glad he's home." This mother's friendly-puppiness shows all the time. She drove me to a museum and parked the car in a place reserved for officials. As we were getting out of the car, a guard came toward us. "You don't mind if I park here, do you?" she asked in the way that she would have asked a friend or neighbor, and taking his assent for granted, she walked away. The guard, looking as though he didn't mind, went back to his post. This friendly puppy is the same with famous people, taking their acceptance of her for granted.

She, and the other incurably friendly people I know, can be listed as successful in terms of their families and their work and the other things we usually judge successful people by. For years, when my life was turbulent and theirs seemed so serene, I thought they had chosen the right way, I the wrong one. I felt that I had been rebellious, not "satisfied as I should have been." There was one thing that I didn't know about my friends: they didn't talk about their troubles as I did about mine. It is only in our later life that they are beginning to reveal themselves to me. I have truly no regrets for my life, no bitterness. It would have been nice to be able to do better than I have, but without where I have been I couldn't be where I am now. When I say this to my friendly puppy friends, they sigh. At other times, they express a bitterness, briefly but recurrently, which makes my toes curl to think of living with it. So it seems to me that I have come off better, and there are intimations that they think so too.

When the eastern therapist writes "pampered" I am not sure what he means, because who means what by "pampered"? He also wrote, "able to lick what confronts him." Is that a true statement, or an ideal? A doctor once told me, "If you can't lick it, run." I ran. Some dragons are too much for any George. In my youth, before I had got so much loused up, and didn't need someone else to tell me what to do, I enjoyed licking what confronted me, but some things looked to me like a steam-roller and it seemed a better idea to step aside than tilt with it. If "able to lick what confronts him" includes being able to accept the fact that there are some things I can't lick, then I agree. Although sometimes it is only for the present that I run away: at a later time I may be able to return and lick what I couldn't earlier. That doesn't mean that I could have done it in the first place. I just hadn't grown up to that one yet.

But what really makes my stomach turn over is this picture of love, love, love, and I love everything you do and never say no. That is what the words say to me. It may be wide of the mark to the man who wrote them.

When my child bites another child, I do not love *that.* But when he explains, "He tore my lei, and I bit him, and he bit me, and (with angry tears) *we each had only one bite* before the teacher made us stop!" I have a deep feeling for him, and love *him.*

When I discover that my child has forged a letter from her father and presented it at school, I do not love *that.* But I recognize that she is in a steam-roller situation and that handling it in this way is necessary, and I love *her.*

My love for my children then is *agape*—not "adoring," but very much being *with.* There is no "pride" (although at other times I am guilty of this). There is humility and humanness. This is what I feel and know and love when I read these paragraphs from Gerald Caplan.*

> We have talked quite a lot about the unfortunate results that may arise from certain pathological pressures in the environment, but these pressures do not necessarily lead to poor mother-child relationships. Even a woman with a neurosis or psychosis may have a very healthy relationship with her child. That may surprise some of you; it did me when I first came across it.
>
> I once treated a woman in Jerusalem who was quite crazy, and had been for years. She had a very bad family life. Her husband was in jail at the time. She was his sixth wife, and I do not think he had gotten rid of some of the others. In any case, he was a well-known brigand. This woman brought her child of six to the clinic be-

* From *Concepts of Mental Health and Consultation: their application in public health and social work,* by Gerald Caplan. Children's Bureau Publication. 373-1959. U.S. Dept. H.E.W.

cause she thought he was stealing. He had taken a pencil from school and she considered this stealing.

Talking with her, we got a depressing picture. She said the child wet the bed and had encopresis; that is to say, he messed himself. And he was very aggressive. But when we examined the child, and we examined him very carefully because we couldn't believe our eyes, he appeared to be perfectly normal. He was a little aggressive and noisy, when compared with a child from Paris or London, but by the standards of his own environment, he was perfectly normal. When we asked him about wetting the bed, he explained first, that his mother could not afford to buy blue paint to put around the door; then, that there were ghosts outside and if one did not paint around the door with blue paint the ghosts came inside; and, finally, that in order to urinate at night he would have to go into the night where the ghosts were. It was very logical that he should wet the bed. As for his encopresis, as we talked more with the woman, we discovered that this was due to the horrible food she gave him. He had diarrhea most of the time, and the hand-me-downs he wore were very difficult to undo. By the time he could unbutton his clothes he would mess himself. Altogether, he seemed a perfectly normal little boy.

She told us that she beat him terribly, and was afraid he would get epilepsy as a result. This sounded pretty bad. I asked, "What do you beat him for? Do you beat him when he wets the bed?" She said: "No! What do you expect him to do? Everyone knows that a child wets the bed." I asked: "Do you beat him when he messes his pants?" She said, "What do you expect? He can't take his buttons off. He's got diarrhea. How would you expect him not to mess his pants. You wouldn't expect me to beat him for that." She also told me he masturbated. I asked, "Do you beat him for that?" She said, "No, that's normal. A little child plays with himself." I asked, "Do you beat him for stealing?" "Oh no," she said, "after all, that's hereditary. His father is a brigand. At the moment he's in jail. I wouldn't beat him for that. Who would beat a child just because his father was a criminal?" Finally I asked, "What do you beat him for?" And she answered, "I have a severe noise in my head, and every now and then I get very nervous because of this. The children make noise. I would hit him with my hand, but I've got rheumatism in my hand, so I hit him with a stick. Only I'm afraid if I hit him with a stick he'll get epilepsy."

When I spoke to the child, he didn't seem to mind. His mother had always beaten him. He said, "When mother gets nervous, she lets fly with a stick."

The interesting thing here is that the mother-child relationship was successful and very healthy. Whatever needs the child had, the woman was fulfilling. She was very warm and protective and she braved the child guidance clinic because the child had stolen a pencil.

Here was a woman with a frank psychosis, living under terrible conditions. Yet this was a healthy mother-child relationship and had produced a healthy boy. There was also a twin sister who seemed very normal. All of this showed us that a woman can have a gross disorder of personality and yet, if her relationship with the child does not involve that part of her personality which is disordered, you may get a child with a healthy personality.

That sounds so different from Carl Rogers—and yet, it isn't. There was an honest relationship between the mother and her child, real understanding and acceptance of the facts of each other—of each other's world as it *is*.

The converse of this would be the parent who is not seriously disordered, but who *is* disordered in a part of his (or her) personality which *does* involve the child. Then, even living in what we consider very good conditions, the child would be in trouble. This happens often.

That statement could make some parents look into themselves with fear and mistrust. "Is there something wrong with me which is affecting my child?" This is likely to make matters worse. It seems to me a better use of intelligence and energy to accept the fact that none of us is in as good shape as we could be, and work on *that.* If I move in the direction of being more acceptant of the child's world as *his,* and accepting as his world what he says of himself, and being honest with him about *my* world and my own limitations (accepting myself), then both of us are getting out of whatever degree of trouble we may be in.

Children and young people are helpful to me in doing this because they are still somewhat aliens in *our* world. As aliens, they are prone to make some mistakes about it, misconstrue some of its happenings in terms of *their* world. But as aliens, too, they are more likely to see the nonsense of some things we have lived with for so long that we don't notice that they don't make sense. They are more questioning than we are. When I join them in their questioning, I make better sense myself.

One child helped me in the de-conditioning of myself simply by asking, of all sorts of things I required of him and of myself, "What's the point?" When I looked into this, usually there wasn't any.

At another time, I welcomed the beautiful and friendly German shepherd, Halekou, who belonged to the neighbors, and chased away an ugly little black dog who slithered along like a snake. A child asked me, "Why do you chase Snakey away?" I said, "Because he's ugly. I don't like him." The child said, "Halekou likes him. Halekou lets Snakey eat from his plate." I was very ashamed, and confused and troubled about how I had come to feel about Snakey as I did. I was touched by the matter-of-factness of the child's observation, and disturbed that I had tried to impose my prejudice on the child. Next time that Snakey came on our land, I called him, and when he came to me I patted him on the head. He wriggled all over with happiness, and his "ugly" little eyes became so bright and dancing that they seemed to shoot out sparks. I shared his joy and loved him.

It seems to me that in education it is more important to question than it is to give answers. This is hardly the way it is gone about in school. I think that "education" is the chief reason why so often when I am asked a question I feel that I must *try* to give an answer, rather than saying "I don't know" or asking another question. Recently when I was being interviewed on tape in an attempt to arrive at something (which we didn't—we threw out the tape) I noticed that when the interviewer asked a question *I struggled to answer it,* when the thought in my head was "That question isn't going to arrive at what you want. Could you ask another question?" If I had thrown out most of the questions this way, we wouldn't have had to throw out the tape.

Another thing about accepting truth is that non-acceptance often brings about a fiction. A Hawaiian woman told me that Hawaiian mythology as it is written in books is all wrong. I asked her how this had happened. She said, "Oh, the *haoles* (white people) asked us about it and then laughed at us for what we told them. So then we told them lies which they believed, and we laughed at *them.*" I have seen this happen with Navajos and anthropologists. One sees it happen all the time within a family. How can it be different in psychotherapy?

How can there be a good interpersonal relationship without respect on both sides? I *don't* respect everyone at all times, but when I don't, it is good for me to know this, and to know that I cannot at such times be helpful to another person. I am fooling myself if I "try to help" someone when I do not feel respect for them. I am not responding to *them,* but to some notion of what I "should" do for people, even though I may truly wish to be helpful. I must not conceal from myself that I'm *not* being helpful.

Most of us would like to be helpful to everyone, and often we wear ourselves out attempting it and then are unhappy about the results. It is good for me to remember Don Jackson's reply to the question "How do you select your patients?" Some other psychotherapists who were asked the question said that they did (or must) take everyone. Don Jackson's reply was, "Life is difficult enough without a psychiatrist requiring of himself that he accept everyone who comes to him." This answer has helped me to be more discriminating in the way that I truly should be, because anything else is dissipation and wasted effort—exhaustion of myself, with no good to anyone else, and the likelihood of harm to both of us. There is no interaction, and without interaction there is no interchange—change taking place in them and me. Each of us continues as before, he in his way and I in mine. We may each become *more* rigid in our separate ways, even though I may have spent many hours "patiently" listening.

In reality, when I think that I am being patient, I'm not. I am "making myself be patient"—putting on the outer manner of patience, acting what I am not feeling. When I *am* patient, I don't notice that I am, or think of myself as patient. It just is. When I *think* that I am being patient, it is much better to look inside me and tell the truth: "I *want* to be patient, but I'm not. I'm feeling all antsy and trying to control myself."

I once heard over Station KPFA Berkeley a talk given by a Negro minister, Rev. Abernathy, to southern Negroes. He really tore into them about their sycophant relations with the southern whites. But he also said, "I *know* I'm being hard on you, but I'm just *not* feeling soft tonight." His words and voice matched, clearly expressing both the "hard" that he was feeling and his regret that he was not feeling "soft." "Tonight" indicated his awareness of change, leaving other possibilities open for the future.

When my honesty begins with *me*, I think it is not harmful to someone else. If it does nothing more, the other has the relief that I heard in a very disturbed person's voice when he said to me, "Well anyway, you're honest."

Wes Westman was a psychology student when he worked in a mental hospital testing patients. He had been doing this for some time when,

> I was assigned to test an attractive woman of about 35, of Italian descent, with a very suspicious nature. She eyed me as I walked with her down the hall to the testing room and as we got to the door she turned and asked accusingly, "Would you like to go to bed with me?" I looked at her and said, "Yes, I think I *would*, but I don't think I *will*." She said, "That's good enough for me. You're the first one around here who's given me a straight answer." She said she had recently asked several men the same question but had not been satisfied with their replies. The testing went well after that incident and she became very open and friendly.

It is better all around if I act from the facts of myself rather than from beliefs about the other—or about myself.

"I am a patient person." "I am an intolerant person." "I am a moral person (with no wayward thoughts or inclinations)." When I decide that "I am" *anything*, this excludes from me *what is going on in me now*, which is the only reality of what "I am." When I am aware of this, I am aware of other realities around me.

> In spite of all similarities, every living situation has, like a newborn child, a new face that has never been before and will never come again. It demands of you a reaction which cannot be prepared beforehand. It demands nothing of what is past. It demands presence, responsibility; it demands you.*

Me. Not what I think about me, or what others think about me, which can never be myself.

With a free interchange between parents and children, there is no telling what may happen. The child who goes to school might well *choose* to go to bed earlier (after a few mistakes) because then he does better and feels better the next day. Anyway, things seem to resolve themselves somehow when I go about it in this way. When my son was eight, he suddenly refused to take baths. It was warm in Hawaii and he perspired a lot. Evening after evening I tussled with him, and then suddenly it made no sense to me to wear myself out that way. So I stopped. And he went to school day after day without a bath. I wondered a little what his teachers might think, but that was between them and him. I didn't expect what happened. After three or four days he came home crying "The kids say I stink!" No more problem.

Quite often the best thing to do is *nothing*. When I was a child, I once stood against the lamp post that was nearest to our house, where I knew that my mother could hear me (besides, it was after dark, and I was afraid to be too far away) and I cried and cried and cried, as loud as I could, for a very long time. I was doing it to punish her, and demanding that she come to me. She didn't come. I went from rage to grief to rage to self-pity while I cried, because she didn't come. She was defeating me. (There were other times when she let me win.) Finally, I gave up because I couldn't

* From *Between Man and Man*, by Martin Buber. Beacon Press, New York, 1955, p. 14.

cry any more, and besides, I knew that I was being phoney, that I was crying *to* punish my mother. I had noticed, too, that when I looked at the street lights through my tears they did fantastically lovely things, like a kaleidoscope, and when I partly closed my eyes, the lights went into crosses. I played with that until I was ready to go home.

I don't extract any rule from that. I don't remember the antecedents to my crying. But I think that my mother did the right thing in these circumstances because what I got out of it was good. I don't know whether she was being "wise." She wasn't very strong and sometimes she let things ride because they were too much for her. Often that was very good for me. I write about this incident because when I have read a lot about sweetness and acceptance I have sometimes felt that I was a terrible parent, and that *I* couldn't conceivably ever have been a good parent, no matter how hard I tried. I don't want other parents to grieve as I did. It does no good. It gets me in a bind. It uses up energy on the past, when all my energies are needed in the present. In fact, it gets me into everything I shouldn't be in, to be a good parent. The Caplan paragraphs are a relief to some young people who don't have children but already are concerned about living with them. One of them gave a shout of joy, and joy is something that we very much need more of. If we had enough, we wouldn't need anything else. I speak of *real* joy, "real" meaning to me that it simply happens, not joy attached to something, not joy dependent on something, but joy that is free. Like the sun, it exists in itself, without purpose—without any purpose that it is aware of—and is a part of and contributor to life.

When a person comes to a therapist, he has a purpose. When the therapist receives the other person, he has a purpose. But when both get into reality together, without facades, the purpose gets lost, and all that is known is something happening that happens of itself, and the noticing of this happening, in you, in me.

Then I am free of all the bits and pieces that I have been holding together, saying "This is me, me, me," and of all the bits and pieces that have been clinging to me by others saying "This is you, you, you." And I am surprised by the power and the tenderness and the movement that comes through as *I*.

VI

> Traditions are a splendid thing; but we should create traditions, not live by them.
>
> Franz Marc

So much of my life cannot be told, and so much that I can tell is hardly worth the saying. There is so much that I have learned not to say because to other people it has seemed bizarre. They became afraid and tried to hide their fear behind a firm manner and confident telling me what to do. If they had spoken their fear, we could have talked about it and my fear too. In hiding it, they gave me something else to be packed away with the things that are not to be mentioned.

I try in many ways to say something about what goes on in me, but no matter which way I say it, it is wrong. I wish that someone would ask a question, express a wish to know. Then I might be able to begin to let out of me what is held so tight inside.

When I was young, and even when I was not quite so young, there were so many experiences which I could not communicate to anyone. One day when I was nineteen, I was walking along a New York street in early morning, feeling good. Not many people were around. The sun was shining and a bright reflection of it caught my eye. It came from a safety pin, lying on the pavement. That safety pin lying on the pavement in a huge city with millions of people and towering buildings had an absurdity about it that made me smile. This switched to seeing the safety pin as a part of the world, as much as I was myself, and I felt myself very strongly in accord with everything, in touch with all there is, known and unknown. Then the safety pin became a key, as thoughts, starting there, went through my head more rapidly than anything that I can think of now, each thought leading to another, like steps building higher and higher, until they arrived at an explanation of the universe. It was all so beautifully clear, and at the same time simple and amazing. I was delighted with this universe in which I lived. Then I thought, "This is crazy. I can't know anything about the universe." Then, what had been clear became a crazy jumble. The speed with which the thoughts followed on each other was too much for me. I fixed my mind on things around me in a determinedly commonplace way—street, cars, people, fronts of buildings—to do what I called "calming me down."

There was no one to whom I could speak about this and other things that happened, and sometimes this made a roaring in my head. Not a real roaring; the feeling of roaring. I held on to that knowing, because if I lost it and let the roar engulf me, I would be lost.

When I had been here on earth for nearly half a century, I worked at a ranch college where there were many freedoms still, and the surrounding country moved me deeply and released me, as it still does. Often I came out of my room in the early morning, walked out on the verandah which was on a rise on one side of the valley, and looked across the fields of alfalfa, the high desert land beyond, and to the mountains on the other side. I would so burgeon with glory that my arms moved to reach to the sky.

I needed the job. I couldn't afford to be thought crazy. I walked down the steps, my arms at my sides, with a singing inside me but a crying too, that my singing couldn't get out.

Behind the bodily response there were many happenings in me full of meaning. I wished to live with them awhile, to know them more deeply, but when I walked down the steps with my arms at my sides, I lost even the knowing that I had had.

We don't have words to express that knowing. Perhaps there can never be such words.

There was one man on this ranch with whom I could converse about these deep meanings and universal understandings that I sometimes felt, by using a few words which he picked up, and with a few words of his own, he gave my meanings back to me so that I knew he understood.

He told me, too, about his world.

"The way I see it," he says, and then there is a long pause, a waiting that is not waiting, a blank space which gets filled in somehow. "It's like a tree . . . ." His arms sweep gently in a way that says something to me about the way that he sees a tree, and the relation of this to what he is expressing about life.

I could repeat all the words he said and they would say no more than "The way I see it . . . . it's like a tree."

My own words say no more. My free gestures, like his, convey nothing to people around us. It is the words, the gestures, the feeling I have from him directly, and the feeling and knowing which they bring from me, which says something.

With most other people, when I try to speak my world, even if they do not reject what I am saying, they insist on having all the blanks filled in by me—clear sentences with precise meanings. When I try to do this, it brings it down to nothing. The life is gone. And then they say, "That's nothing." I have presented them with a cadaver, not a living being.

I feel that I have desecrated life, and I am sad.

I feel cut off. I long for my schizoid friend who does not require of me what is not possible, and who accepts with understanding what I say. We stand side by side or sit on the same rock, not touching each other, but the boundaries are lost without touching and I feel that we are one, and the more that I am with him, the less inclined I am to "carry on a conversation" except in terms of utility—explaining to another person something that needs to be done, or what I need him to bring to me. For utility, this language is good. I can explain why a nut should be put back on a bolt when the bolt has been removed from whatever it was in. I can answer a question about the best way to get back to town.

But for being with people, what I love is the poetry of life, the art, and the language which conveys this, in which there is a vast expanse, something not put in, which my large mind can encompass. My *small* mind cannot understand that. It wasn't built to. My large mind cannot learn to fix dinner. That's not what it is for.

When I am free, my small mind and my large mind function together. I don't know how they do it. But it's something like breathing, which I say that I am doing without usually noticing that it has two parts. I do not think "I am breathing in" "I am breathing out" unless for some reason I am paying attention to it. Otherwise, I am just breathing. When I say "My breath stopped" or "I caught my breath" I do not think whether the breath was out or in.

It's something like that with my large mind and my small one. Not much, but something like: at least, *both* are necessary.

If I am too much in my large mind, I lose touch with earth. If I am too much in my small mind, I lose touch with heaven. Not a hereafter heaven, but a heaven *now*. When I am living with my bothness properly, my work is joy, my play is joy. When I am living with my bothness properly, life and death do not concern me: I am prepared for both.

When for long periods of time I am never with anyone who speaks the language of art and poetry and music, or with the rest of nature, more and more I feel tight and bound, isolated, living in a world where I don't belong, where all the people seem to be against me. I can easily feel that they are my enemies, because they get between me and my joy.

If people push me down into a cellar and shut me in a room without light or air or sun, are they not my enemies?

And that is what people do to me so often.

When I do not confuse the image with physical reality, I can feel that this is what they are doing to me in the physical sense. So long as I know that it is only the image expressing to me in the outside way what is going on inside me, I am all right, even though I hurt and am deeply distressed and feel trapped, as though people won't let me out.

Am I really wrong that people won't let me out?

I want to share my world with them, to have them roam with me and freely play and dance and sometimes weep, but they insist that I come into their world which is "real."

Both worlds are real and one when I do not split them.

Sometimes when I have gone into their world, the half-world that is called real, the pain is so great. . . . I don't know how to live with it. There is horror, too, at the unknowing cruelty of people, like animals tearing a person apart because to them a person is just meat.

Then I feel like shutting my small mind which sees only the small world that is known as *this* one, and curling up in a ball with the half-world inside me, known as *that* world, and making it my whole. Or else I feel like lashing out at the beasts who do not know I am a *person,* trying to get from them what they have taken away from me.

People are afraid of my *whole* world, afraid that I may get lost in it and have no conscience and no caring. But it is in the *whole* world that no one can get lost. In either half-world I am lost because I have lost the other. I think that I have made my choice, have chosen my whole.

But in this there is no choosing, any more than I can choose between night and day. My whole is both, and my happiness cannot *be* without it. I am always searching in one half-world or the other for what I need to complete me, to make me whole.

*Subverbal Communication and Therapist Expressivity: Trends in Client-centered Therapy with Schizophrenics**

Eugene T. Gendlin

Introduction

Growing out of client-centered therapy and especially its application with hospitalized schizophrenics, there is now emerging a mode of psychotherapy which centers on the "experiencing" of the two persons in the therapy interaction, rather than on the discussed verbal contents. This development is similar to recent trends in other orientations: there is a strong tendency to emphasize the interaction in psychotherapy, to emphasize that two human beings are involved, and to focus on the concrete subjective events occurring in these two persons, rather than only on the verbal contents being discussed.

Basic to this development is the view that psychotherapy involves "experiencing" (by whatever name), a somatic inwardly felt process, the manner and meanings of which are affected by the interaction. Some attention is being given to the theoretical problem of how inwardly felt bodily events can have "meaning," be "explored" and "symbolized," and how these concrete implicit "meanings" can be affected and changed by interaction. Now that psychotherapy is widely thought to involve a concrete feeling process, we are less specific about the (still vital) role of cognitive symbols and

* Reprinted from *Journal of Existential Psychiatry*, vol. IV, no. 14, 1963; copyright 1963, Libra Publishers, Inc. The first five pages have been abridged here.

exploration. The different orientations use different cognitive vocabularies, yet their patients and clients appear able to work with any of these vocabularies. Apparently, any good vocabulary can be employed as a symbolic tool for "working through" and interacting. It appears that personality difficulties lie in the "pre-conceptual" meanings of experiencing—that they are concrete, and that they are amenable to symbolization and change through interaction using *any* of the many different therapeutic vocabularies we have.

Apparently it is not so much a question of just which vocabulary you use, but rather *how* you use it. If used in "direct reference" to experiencing, then more or less any vocabulary can be well employed. If used as abstract explanatory substitutes for the individual's experiencing, no cognitive vocabulary will engender much constructive personality change. Of course this is an opinion which awaits much more research support than we have so far, but it is a widespread current trend in therapeutic thinking.

Not only the best use of words, but also the best use of therapist behaviors, would seem to depend not so much upon just what the therapist behavior is, as on *how* the behavior relates to and affects the individual's experiencing. The attempt to specify just what therapists should and should not do tends today to concern not so much *what* behavior is used, but more just *how* it is used. Furthermore, this "how" concerns how the behavior relates to the experiencing of the two persons.

Purpose of this Paper

The purpose of this paper is to offer a few specifications of just *how* subverbal communication and therapist expressivity can work. Such specifications are needed because, if therapists take it as a rule to "express" anything whatever, in any way whatever, obviously we are left with no guidelines at all. The specifications will, however, concern not *what* but *how* such therapist self-expression can be part of a therapeutic method. Similarly, "subverbal" communication always sounds mysterious—how is anyone to know what the client means when he can't say what it is? Again, the specifications I would like to offer concern not "what" the client means (and not how we can guess "what" he means). The specifications concern how client's and therapist's words can be employed to point at and refer to experiencing.

Schizophrenics and Neurotics—some research findings

Although psychotherapy seems to me to be *the same* process in neurotics and in schizophrenics, the differences manifested by schizophrenics are important, because here we find written large and unavoidable some of the factors which we may overlook in usual psychotherapy. For example, we often tend to overlook the problems of externalization and trivial talking in ordinary therapy. We know it happens and that such clients are most often eventual failures, but how different would be our reaction to this problem if nearly *every* person we saw in therapy confronted us with it. We would then *have* to do something to meet the problem.

We do not as yet have conclusive findings about the effectiveness of psychotherapy with schizophrenics. Therefore, what I have to say will not be evaluative. Rather I will be describing what we are actually doing in our therapy sessions and how we have come to proceed as we now do.

Therapist Expressivity

Rogers, in 1957, posited three conditions necessary and sufficient for psychotherapy. They are: "empathy," "unconditional regard," and "congruence" or "genuineness." This last, "congruence," implies that the therapist attempts to drop any personal or professional artificiality, any maneuvers or postures, and that he be himself. In our work with schizophrenics this condition has become more and more important. Gone are formulas—even that most characteristic of client-centered modes of responding, which was called "reflection of feeling." As the term "empathy" implies, we strive as always to understand and sense the client's feeling from his own inward frame of reference, but now we have a wider scope of different behaviors with which therapists respond to clients. In fact, I believe that it was in part the undesirable tendency toward formulas and stereotyped ways of responding which perhaps led Rogers to formulate this condition of "congruence" as essential.

To 'be himself" has also meant that the therapist has become more expressive. The therapist much more often expresses his own feelings, his experiencing of the moment. When the client expresses himself, then naturally the therapist's present experiencing consists largely of an empathic sense of the client's meaning. But, when the client offers no self-expression, the therapist's momentary experiencing is not empty. At every moment there occur a great many feelings and events in the therapist. Most of these concern the client and the present moment. The therapist need not wait passively till the client expresses something intimate or therapeutically relevant. Instead, he can draw on his own momentary experiencing and find there an ever present reservoir from which he can draw, and with which he can initiate, deepen, and carry on therapeutic interaction even with an unmotivated, silent, or externalized person.

Also, "congruence" for the therapist means that he need not always appear in a good light, always understanding, wise, or strong. I find that, on occasion, I can be quite visibly stupid, have done the wrong thing, made a fool of myself. I can let these sides of me be visible when they have occurred in the interaction. The therapist's being himself and expressing himself openly frees us of many encumbrances and artificialities, and makes it possible for the schizophrenic (or any client) to come in touch with another human being as directly as possible.

However, it would seem that many of the erstwhile guidelines for therapeutic *behavior* have disappeared. Only the *basic attitudes* are specified, not *how* one manifests or expresses them. Does this mean that we do just anything at all? Let me describe this procedure of therapist expressiveness in more detail. I will specify it further in three respects:

(1) *"Non-imposition":* In working with unusually defensive, withdrawn or fearful individuals we find it even more important than ever not to impose ourselves on them. How is non-imposition consistent with a therapist who expresses more of himself more openly and actively, and who initiates relationships by expressing himself? Tentatively, I think the answer is: the therapist can be more active and *at the same time* present less imposition and threat than ever, if he will express *himself*—his own imaginings, his own feelings, desires, the events which transpire in him—and if he does this clearly

and explicitly as statements about himself, or about events transpiring within him at the moment. In this way he shares himself more openly, yet he does not impose on the client's experience. He speaks for himself. He does not impose or force anything upon the client's experiential space, or confound events in himself with events in the client.

(2) *"A few moments of therapist self-attention":* To respond truly from within me I must, of course, pay some attention to what is going on within me. As I am interacting with the client, a good deal of what transpires within me has to do with him—consists of my imaginings of him, my observations of his reactions, my reactions to him, etc. But, within me these occur as mine, as me. They are not deductions about him. They are what is happening to me now, my lived moments with him. To formulate and express these I require a few steps of self-attention, a few moments in which I attend to what I feel. Then I usually find a great deal that I am willing to share. It would be wrong to say that I express *everything* that is going on in me, since a thousand things are going on in me at any moment and they cannot even be separately formulated, much less expressed. Also, I don't blurt out impulsively the first thing which happens to come to mind. I live a few moments inwardly and by this means I find in myself some response to the client, or to what has been happening between us, or to our silence. Even when little is being said, I find that I have desires, fears, disappointments, and wishes for more meaningful communication. I can voice these. With a few moments of self-attention I can find my genuine response to the moment, If, while the client talks, I feel bored, I do not blurt out "*You* bore me." I find with a few seconds of attention to my own experiencing, that my boredom really consists of my missing something from him, something interesting and personal. I find that I strongly wish for this personal expression of his. I find I have much welcoming for this personal expression ready, and that it is going to waste. I find that I can imagine the kind of personal communication which I sense is missing in his verbal stream. I can express these senses of lack, wish, and imagination, and I can express them as mine. Much of my own feeling process when I am with someone does usually consist of these rather specific momentary events, reactions, wishes, and senses relevant to the other person. For example, let us say I have just

said something and gotten no response. I think that it may have been very much the wrong thing to say. I need not simply feel bad about having done the wrong thing. I can say I feel bad about it and why, as well as the fact that this happens to be what I am now feeling, but, that I am not at all sure about what *he* is now feeling.

The few seconds of self-attention nearly always bring two developments in the feeling I have: (1) it becomes more truly something of me, rather than about him; (2) it becomes much more possible to share it. Thus, even though it is my genuine momentary reaction *to the client* in the interactive moment, it is also *genuinely mine* and does not impose itself on what he experiences. I can say—when it is true—that I am not at all sure what he feels at the moment.

Thus, the two specifications I have stated require each other: "non-imposition" requires "a few moments of self-attention" so that I may find what I truly feel and state it non-imposingly, as mine.

(3) *"Unmuddled simplicity":* I want to now add a third specification of therapist expressiveness. When the client gives me nothing intimate, or self-expressive—*it is then* that I must draw on my own momentary reservoir to find within me my response to him—a clear, intimate expression of me in this moment with him. However, when the client is in the process of expressing himself to me, then I find within me chiefly my sense of *his* expressions and I try to tell him with unmuddied simplicity what I understand *him* to feel and think. It is just as important that I sense and state his experience as *his* (as purely as possible from his own frame of reference), as it is to make clear that my experiential self-expressions are *mine,* when that is the case.

Very often, when the client is in the course of expressing himself, a response that simply states what the therapist understands the client to think or feel is a powerfully effective response. Often it is the only possible helpful response.

I have described three specifications of therapist expressivity: "non-imposition," that he states his self-expressions as his own; "a few moments of therapist self-attention," enabling him to find his true response of the moment; "unmuddied simplicity," in stating the client's feeling or thought when the client is expressing himself and the therapist finds within himself chiefly his sense of the client's message.

With schizophrenics and with many others, it is the therapist who—if anyone does—will initiate the relationship, will begin open and expressive interaction, and will first express warmth, care, interest, and a person-to-person quality. If the therapist must sit passively, or argue intrusively, I don't think he is likely to form a relationship with an individual who does not already wish for therapy or for a relationship. The therapist's moment-to-moment expressiveness largely determines the quality of the interaction, at least at first, and especially with unmotivated individuals. The therapist's self-expression can make the interaction eventful, personal, and expressive even if the client is consistently silent or expresses only trivia. The therapist's expressions—the events occurring in himself—will concern the interaction and will deepen it, when they are spoken. *Both* persons tend to experience an eventful, open and personal *interaction* even while only one of them is verbalizing his felt side of the interaction.

This brings me to the second observation I want to discuss:

Subverbal Interaction

Perhaps subverbal interaction is so important with schizophrenics because so much of their experience seems incommunicable to them, seems cut off from other people in its very nature. Often, the content—the *what* is *said*—is only a small bit, perhaps a bizarre bit, which issues from an inner turmoil whose incommunicable significance is enormously greater than and different from any bit of verbal content. The incommunicable nature of *what* the individual experiences, and the fact that his experiencing is cut off from other people, require that one respond not to some bits of verbal content, but to the experiencing. In this way one attempts to restore the connection, the interpersonal *interaction* process within which the normally functioning individual lives and feels.

This is not to say that one mysteriously responds to experiencing without having some verbalization to go on. Rather, one views verbalization differently. Instead of concerning oneself with its content one asks: What larger inward process is this bit of verbalization coming from? One's answer to this question will be something felt, a conceptually vague but concrete felt meaning which the client feels and thinks, and which the therapist can only imagine. But the therapist need not know it, guess it, or correctly imagine it. He can point his response at it, no matter how unknown it is to him.

For example, my client says that he wants to know where, in the hospital, they keep that electronic machine which compels people to return to the hospital. He can prove there is such a machine, he says, because how else can you explain the fact that patients with ground privileges return to the hospital of their own accord?

Now, I could, of course, argue with him that no such machine exists, that I would know if it did, that he does not trust me to tell him the truth on that subject, that he is having an unrealisic hallucination, or, closer to his feelings, that he does not like the hospital and cannot understand anyone's coming to it voluntarily. But what is his *experiencing* as he talks of this machine? What is the "pre-conceptual" or "felt" meaning from which this bit of bizarre verbalization comes? I don't know, of course. But I want to respond to it somehow. So, I say back to him: "You have felt in yourself the effects of this machine you are talking about?" "Of course, I sure have," he says and goes on to say that the machine makes him feel "not himself." This phrase I recognize as somehow communicating to me something of the inward experiencing at which I pointed my words.

I am using this example to illustrate what I mean by pointing one's words at the experiencing, the wider inward process about which one does not know very much—except that it is there—and that verbalizations arise from it (or in regard to it). Actually, I had not quite correctly imagined what he experienced. For my part, I thought he felt compelled inwardly but his next words expressed a slightly unexpected but still understandable aspect of his experiencing. And this is what usually happens. Usually, when one points one's words at the experiencing rather than the verbal content, one finds that one's imagination was not accurate, but the very fact of responding to this everpresent experiencing establishes the possibility of communicating concerning the deeper meanings from which the verbalizations arise.

This man went on to tell me that his "not feeling like himself" was the result of the fact that his parents moved into the country when he was a schoolboy, and that therefore he had to ride a bus to school through the snow many miles. Again, one might have argued that this alone could not have caused him to feel "not himself." But one senses that this bit of memory comes from a whole pageant of memories, and, since it involved "not being himself," I imagine endless, weird, snowy bus rides, I sense his feeling cut off from everyone he knew, way out there, snowbound, in the country, those many years all of which he now feels, I suppose. I say something about these bus rides and feeling cut off and we establish a new vehicle of communication. He too now uses the phrase "feeling cut off." Perhaps I was right—but more importantly, I spoke to that mass of felt meanings and thoughts, that feeling process, which was just then occurring in him as he spoke, not to what he spoke as a bit of verbal communication. And, in this way, although often very stumblingly, one can gradually communicate more meaningfully, despite bizarre or externalized and trivial verbalization.

But I have chosen easy examples. Before this man told me about the electronic compelling machine, we spent six hours together with only trivia and silences. It was necessary for me to respond to him when, as yet, he would share nearly nothing with me. What was I trying to do with him, he wanted to know, and when would I be through? When would he not have to come any more? When could he go home? He had nothing to say. Silence, more silence. Once I interrupted one of these silences in which he had been sitting very quietly, apparently thinking, and said very gently: "You seem to be thinking some important thoughts or feelings. I don't know, of course, but that's what I imagine. I don't want to interrupt but I sure would like it if you felt like sharing those thoughts with me." He said, very loudly: "What? Who, me? What, thinking what?" It was quite clear that he was startled. Also, he seemed to consider my statement inappropriate, false, and stupid. Yet, it is necessary to bear such moments, for how else can our interaction come to be warm and close and personal, if one of us doesn't make it so?

After a while such expressions of mine, such imaginings, or implications that we both experience important feelings, are no longer met by surprised rejection but often by a silence that is close to assent; and then, later, by an explicit sense on the part of the client that our silent times are subverbally important, deep, and eventful. One client named it when she said: "I'm having quiet therapy for a while."

An even better description of the inward feeling process which can occur in this subverbal interaction came later, when she could describe it much more specifically. She says: "When I get upset, I can't breathe. I know I must be breathing, of course, but it feels like I can't. Here, after I've been here a little while, I breathe."

She means, I think, that in these silences there is an inward flow, an inward feeling process which comes alive or is released.

I do not mean to say that subverbal interaction occurs only during silence, although silence appears to be important for it. In my first examples which came from a conversation, I tried to show that, while we observe silence and words (and the words before and after a silence) very much more is occurring within the person in his feeling process. Often, what he says and how he looks gives us a little to go on in responding to this feeling process. But even when there is nothing to go on, one can still respond to this process, point one's words at it, express something from our own felt process, thereby engendering a deeply important subverbal interaction.

Subverbal interaction in one extreme might be illustrated by the man with whom I spent nearly six months of bi-weekly meetings which consisted of my standing next to him on the ward. Whenever I asked him to come into the office with me he became verbal, but only to tell me to go away and leave him alone. Yet, when I would come only to stand next to him where he stood on the ward, he would usually remain for the whole hour despite the fact that he knew I would leave if he walked away. During such an hour we would exchange many glances, motions, and a few sentences. Often—that is to say, every few silent minutes—I might say something of the tension I felt, and of my wish that our silences would be or feel all right, as well as of my wish to hear from him, and my knowledge of his discomfort and tension with me. After a while I did usually receive a sentence or two, often in the nature of summaries that seemed to stem from much ongoing inner turmoil, feeling, and thinking, such as: "Maybe I *am* crazy," or, "someone must a have a use for a person," or, "I don't know if you're for me or against me," or, "they *just* don't have a heart," or, "I'd like to take them by their shoulders and shake them to wake them up." Some hours passed without such expressions. Sometimes he would accept my responses to these statements, more often he would show me that I could more easily be talked to if I was quiet: he would say, "don't pressure me," or, "you're too curious," or, "maybe you *are* against me," or, "it's awful hot today." At slight movements, or unsureness, or verbiage from me, he would react with sudden black looks or leap away three feet. So I learned to be *quiet while* he talked; perhaps a few minutes *later* I might tell him something of what I thought about it.

When I stood next to this man in silence, it was not the case that nothing happened when we were silent. Clearly, he was very active inside himself, and to me it was also clear that I had a great deal to do with the process, and with the quality of this process, within him. It is this kind of interaction which I term "subverbal."

The development from early to later interviews involves the establishment of subverbal interaction. In our research, the proportions of speech and silence do not change from second to thirtieth interviews. However, not only do the silences come to be subverbally important and therapeutic, but the verbalizations also become more significant, as the findings on the Kirtner and Experiencing Scales show. Subverbal interaction is thus not a giving up of verbal therapy, but rather a reaching for the deeper and wider feeling process which occurs in every individual at every moment, and within which psychotherapy occurs. Words, no matter how relevant or irrelevant, are only messages from this deeper feeling process, only symbolizations of experiencing.

VII

> People often take prejudice or habit for truth and in that case feel no discomfort, but if they once realize that their truth is nonsense, the game is up. From then onwards it is only by force that a man can be compelled to do what he considers absurd.
>
> Alexander Herzen (*circa 1850*)

Each time that I read this paper by Eugene Gendlin, I feel so deeply *with* it, so much wanting this *for* me and around me, and the more I study it and try to practice what he describes, the more meaningful it becomes to me. The happiness that I knew in living with Hawaiians, Navajos and Hopis begins to seem possible for us white folks, too. Is there anything in the three basic attitudes that doesn't *belong* in our daily lives? When I have lived this way with other people, the *content* of conversations was different, but the *approach* was so very much the same. When I live with people in this way, I become a better person, and also a happier one.

"But *we* don't have time. The complexity of modern life . . ."

"Why not slow down?" I ask.

A look of exasperation at my unworldly imaginings, my inability to see what is required of people today because it's "necessary." I am an old lady with wool in her head.

But it was the Chief of Police in San Francisco who said, when asked what could be done to decrease the number of suicides, "We gotta slow down."

Why should we arrange a "civilization" which is a torture to many and not very good for anybody? The world we live in doesn't *have to be* the way it is. Out of all the possibilities, we chose to make it this way. Those who never did like it helped to make it by going along with it. What people don't go along with falls apart, so what *is,* people are going along with. Even a military government loses its power when people don't go along with it. In Hawaii after the Japanese attack on Pearl Harbor, we had martial law, and one of the first acts of the military was to make it unlawful to speak against the military. But everyone was *so* mad at the officers of the Army and the Navy at that time that they went on talking everywhere—in the streets, in restaurants, grocery stores and hotels—and what could the military do? I didn't hear of anyone being even rebuked for speaking against the military. You can't punish *everyone.* Physically it isn't possible and besides, then who is left to work with you? Leaders are non-existent without followers.

The number of people who now don't like what we have made seems to include almost everyone. But we're all split up in groups which don't like this and don't like that, who are fighting each other, not knowing that what we deeply want is something else. It seems to me that this is a common cause of juvenile delinquency and other forms of mental illness. I include in "mental illness" the madness for *things,* and treating people as *things* (however nicely) beginning with myself who has made a thing of me. In our bewildered desperation, we can only think of "fun things" to relieve us.

Connie Fox, an artist whose paintings are increasingly pre-conceptual, writes:

> I think you present the idea of sub- or non-verbal communication very well. It's fascinating to me to listen to someone else's non-verbal world. Its existence is very familiar to me. I mean it's a familiar world. But I wonder about the "words" you use—not the meaning, for I believe I understand that—where you write about people making "things" so important. But things can be loved and remain "unchanged." I remember someone asking me about the Volvo, did I like it, and I said, "Yes. I find it more responsive than most people." And I don't consider this a withdrawal from people, but another kind of love, as for landscape or country (cultural, not patriotic) or built things (the love expended

on the Golden Gate Bridge)—a sustaining kind of love, like work can be sustaining.

"Like work can be sustaining"—the problem of *words* again, for of course I can be sustained by work if I am using it to fill the void that has come about through my inability to love, or the inability of those around me to love which cuts me off from my love. *Or,* I may be sustained by my work because I have done it with love. As Gendlin says of therapy, "not *what,* but *how.*" There are two ways to ride a horse (or drive a car, or clean house, or take care of a garden, or wash dishes, or teach, or build, or anything else). One rider demands that the horse be obedient to him, he makes the horse a "thing." If he has any feeling for the horse, it is the feeling of possession, of "mastery." The other kind of rider is, with his horse, more like a centaur—horse and rider move together, responding to each other in a way that makes them move as one, they *understand* each other. When I can live with people in this way—and there *are* some people with whom I can do this most of the time—there is joy, and what is to me the peace that passeth all misunderstanding. I myself become a "centaur"—body and mind move as one, not "myself" making my body do what "I" (or others) want it to do.

It seems to me that this is the way that we are built to be, moving as one with my body, as one with others. What we call "work" or "being practical" has had a lot to do with pushing us out of this. At this time, when cybernation is taking over so much of the work and is capable of taking over so much more, why not let it? Why should I be tired in a country which can turn out too much of everything? Why should I not rest when I am tired—for whatever time it takes until I feel rested? Why should my body be overruled by me? At this time, we have a chance to be human and discover what "human" is. If we don't do this voluntarily, we may even be forced to be human because there's nothing else left for most of us to do.

What is it that I want if not release to a feeling of ease, of completeness, and a release of the talent that I know I have for responding to all the world, including people? Is there anything I want more than to feel fully alive and *living?*

One afternoon I listened to a tape of Carl Rogers in therapy with a man who thought he was "no good," and because of this he wished he were dead. Carl accepted that and did not try to change it. There were very long pauses. Carl said once, without the emphasis that is imposition, "I know that you don't care (if you die) but I do"—a simple statement permitting the other man to be as *he* was, while at the same time conveying, "but you do mean something to me," or "to me you are not worthless." At another time, Carl expressed his knowing what it feels like to feel "no good" by saying that he had sometimes felt that way himself. He expressed his understanding of how the man felt, too, by saying, "you just feel no good—no good at all," with his voice conveying what it feels like to feel that way. The man sobbed. He sobbed so long. At the end, Carl offered him some kleenex and told him where he could wash his face, so *neutrally* that there was no switch, no checking of the movement going on, in them, in me.

All of me was alive with movement, with sadness and joy intermingling without conflict—sadness that *anyone* should feel so utterly "no good," that fellow humans should have made him feel this way, and joy for the release of his sobbing, and for the person of Carl who released this, through his own humanness. I was feeling very human myself, and wanting nothing to interfere with that. I wanted to live with it awhile. I couldn't listen to another tape, and wanted to go home without having to talk to people superficially, which is to say unhumanly, on the way. I started to go quickly. And then I noticed a light coming through Carl's office door. It showed on the hall carpet. I was moved to go to him—my body moved a little in that direction. But then the *I* which is *not I* thought, "No, don't. He's a busy man. You can tell him some other time." But *I* came through with the knowing that no other time would be the same. It couldn't. If I told him later, I would be trying to recapture something, to make a copy of it, reproduce it, when it was the living *now* that he had produced in me which I wished to share with him. I erased the conventional "do not disturb" thought from my mind. That's all. When I did that, what had been checked for a moment moved, and I moved toward his door without thought or intention—both of them had disappeared when I erased the admonition "don't."

Carl was sitting at his desk, wearing a white cotton shirt. I told him I had heard the tape. I tried to "think of something to say" but I couldn't. Words wouldn't come into my head. My vocal cords felt stuck together. But then something said itself—"It's beautiful!" His face expressed happiness, but I still felt "unsaid," that I hadn't altogether conveyed me. But no more words would come. I wished to touch him, to communicate in this way. A wispy thought came saying "No." ("One doesn't do that.") When I erased the thought, my hand moved out and touched his shoulder. His response to me was as clear as my response to him—to both of us—although all he said was "Thank you" and all that I did was leave.

My love (*agape*) for Carl was never more clear to me than in those few moments in his office, and there was something very much going on between us. *This* love is the love that is *real* to me. In it, there are no demands and no necessities. There is no image that I carry around in my head and idolize, and get cross or hurt when this image which I have dreamed does not live up to what I have dreamed it to be. This love to me is real because it accepts the reality of the other person as he *is*, not wishing him to be something else. This love does not prohibit sex, but it is not sexual. There are men, women and children whom I love in the same way. It includes all ages.

I do not *want* anyone to love me exclusively. That pinches me. It binds me, demands that my love flow in one direction, when my *love* is free. When it is not free, it is not love to me. "You are *mine*," gives me the shudders. I wish to be free myself, and to let others be free of me.

The *efficiency* of largely non-verbal communication in interpersonal relations is prodigious. How much of a busy man's time had I used up? And when I left, I felt that there was nothing, no least bit of what I was feeling at that time—including my awkwardness—that had not been conveyed. I felt that I had been totally received and understood. When I left, it was not because "he had no time for me" but because that was all the time I needed. Everything had been "said." It's like the kind of poetry which conveys so much in so few words, with so much emptiness, that when you try to analyze it, you can't. "Either you get it or you don't."

Sometimes I have said so few words that the meaning could

easily be misconstrued, but when the other person is receiving, she catches what I have not said. In a group, a therapist was distressed because he could not get through to a woman. He made no attempt to conceal this, and to me it was clearly *real*, that he *wanted* to be in communication with her. But to her, he did not express pain: he was being "phoney," putting on an act "to break her down." I said to her, "Perhaps it's a different way of taking pain." This happened to get through to her, as words usually don't, and she asked quickly, "What do you mean?" I said, "This isn't *you*, Mary, but you *seem* to be a cream puff. . . ." I made a motion of floating along, with my hand. She responded so completely to everything that I thought and felt that she let go her control of the person she presents herself to be, the role, the disguise, the shell. The mask dissolved, her face crumpled, and real Mary looked at me from eyes that were filled with the pain she never shows.

When I heard my words alone, after I had said them, hearing them again in my head, I was astonished. Surely I must have said more than that? But I hadn't. Even as I said the words, they weren't making much sense to usual-I. There was all sorts of stuff going on in me at the time the words were spoken which made their meaning clear, but *this was not said*. Yet Mary received it.

To accept any credit or appreciation for what happened would be like saying "How clever am I! I'm breathing!"

It's like my daughter when in art class "she" painted something so unlike and beyond her usual work that the instructor looked at it and asked, "Who did that?" My daughter replied, "I don't know. Some ghost, I guess." What would life be like if we trusted our "ghosts" all the time?

When I had been so very sick and miserable and both broke and broken for a year, the doctor said one day with deep feeling, "I'm so glad this happened to you and not someone else!" Then he looked shocked, as he heard only the words he had said. I laughed because what he *meant* was as clear to me as what he had said, as clear as his astonishment and shock that he had said it, and everything that was behind all this. He was thanking me for being me, not someone else.

Is there anyone who hasn't said something at one time (most likely more) and then realized that the words said the opposite of what he meant, and yet within himself the meaning and the reason for the words was completely clear? Usually, when a person hasn't understood me, then I give up, because the more I explain the more sure he will be that I am apologizing, trying to be nice after having said something that I felt but didn't mean to say.

This doctor and I went through some very troubling times together. At one time in the hospital, I wanted to give up. It seemed so much easier to die. He didn't let me. At another time, he lost confidence in himself and offered to turn me over to any other doctor whom I chose. I didn't let him. Between us there was *agape*—and mutual understanding. This love was not always tenderly expressed as it was one day when I said that I had never known an angel and he said, so low that I almost didn't catch it, "I know someone who is *almost* an angel," but it was just as strongly present on the day I said, "The muscles in my legs seem to be loosening up. Is that a good sign?" and he looked at me as though I should know better by then than to ask questions that he couldn't answer and said, "That's just *rigor mortis* wearing off." I had been through so many somersaults that I wasn't certain that I hadn't died, which made it funnier to me than it could be to anyone else. It was like not knowing which house one is in because it has some of the features of both.

When he said, sounding cross, "I wish that you'd cooperate!" I knew completely what he meant, and why he sounded cross.

And when I answered, "You don't give me anything to cooperate *with!*" he said "Ouch!"—but knowing that I knew that he was giving me everything he could and that we both felt there was something we were missing out on.

There was no sticky sweetness in tenderness like the "angel" bit, no nastiness in the *rigor mortis* answer or in either of us in zipping back and forth about "cooperation."

There was suffering and struggle, self-doubt and frustration for both of us in those years when we were trying to get me well, but when I was leaving to move to another town he said, "In a way, it's been fun." I thought so, too.

VIII

> The new insights for which I am trying to find a place have mostly been suggested by recent experiences in science which I believe have revolutionary implications not appreciated even by most scientists. Two convictions have been growing upon me—a conviction of the importance of a better understanding of the nature and the limitations of our intellectual tools, and a conviction that there is some fundamental ineptness in the way that all of us handle our minds. . . . Neglect of the role of the individual, with resulting overemphasis on the social, may well be one of the fundamental difficulties in the way the human race handles its mind.
>
> P. W. Bridgman*

When I move into other societies, or even sub-cultures within my own society, I am often astonished by their tabus. I got into one group of people this past year who live in a world that I have pretty much kept myself away from, out of a feeling of not belonging, not being in accord with, having different values. Several people in the group were visibly and audibly shocked when I referred to myself as an old lady.

It is only very recently that I have begun to see *as tabus* those experiences which in my own society must not be talked about, and myself a victim of the edicts of the gods of Science and Reason that have been set up in the western world, superseding all other gods. It seems to me neither

* From *The Way Things Are*, by P. W. Bridgman. Viking Press, New York, 1961, p. 1.

scientific nor reasonable to exclude any human experience, anything that occurs to any me. But that's what I have been doing. I have been excluding not only from others but from *me* those experiences of my life which are tabu. I told myself they didn't happen, and so lost touch with them myself. During the past seven years I have increasingly found these experiences to be acceptable—although certainly not to everyone—and this acceptance has made it possible for me to accept them in myself and live with them. My life feels much more alive and lively and more free.

Interpretation of experience is another matter. It is easy to go haywire on that, particularly if I insist on *one* interpretation. When I was sick, my sickness was diagnosed "organic"—the somatic part of psychosomatic—but I didn't feel sure about that, so I studied myself. I discovered that while I could *decide* for either psychic or somatic, as soon as I removed my decision there were *both*—interacting now as they had through all my life.

Next, I asked me if I had brought my sickness upon myself (as some people said). I went into this deeply and found that I could build a good case for it, and that I could build an equally good case for my sickness having been dealt out to me as a punishment (to make me learn). Or it could be retribution. Or simply cause and effect. In a groping, unclear way, I wrote to Aldous Huxley "Perhaps it is all of them and they are all the same thing seen differently." My difficulty in accepting the truth of my own discovery was purely and simply that I didn't know anyone else who saw it as I did. Each person had *one* view, and that was the way that I was supposed to see it. When I said (experimentally) that my sickness was organic, some people got mad. When I said that it was psychosomatic, others got mad. When I said that it was both, everyone got mad. The effect of everyone being mad at me was to make me feel that I must be "mad" myself. When I tried to express my stumbling beginning knowing that the *same* experience can be seen in many ways and expressed in different concepts or theories, any of which may be useful to me if I like it and use it wisely, then I was "mad" to *other* people. "Barry's smart, but if she thinks she knows more than the experts, she's crazy."

Which experts? At that time I didn't know the ones who would agree with me. I hunted them out, over a period of years, to prove to me that I wasn't crazy, because when I thought that I was, I *became* a little crazy—mixed up and confused by the difference between what I saw and what everyone else around me was seeing. I am no longer troubled about being "crazy"—for probably the first time in my life—because I know that I can become confused, but also I can become unconfused. I can let myself go into confusion with confidence that if I let myself go along with it, I'll come out all right. I am no longer afraid that this in itself may be crazy—to let myself go into a kind of madness, to explore it—no matter how many other people may think that it is. Without the fear, there is nothing to be afraid of. I can accept my own experience and know it to be *me*.

At the time when I was in such great confusion, I was also having difficulty with the doctor, although this was in a friendly way and he didn't think that I was mad. I had had some notion that the doctor knew, from the observable symptoms, what was going on in me. I found that he didn't know at all, and that there was a big discrepancy in our two views. He was interested in mine, as possibly representing what went on in his other patients that he didn't know about and so did not understand *them*. I was interested, at first, in trying to "correct" my view to his. But that didn't work out. My view *was*, just as surely as his view *was*. So then I tried to find some way to fuse the two views. I tried to make them come together, with some cockeyed notion that there "should" be only one view. I exhausted myself in trying and gave up. When I gave up, I saw both views as *different*. I didn't have to do anything about it—no trying to force a reconciliation. When I let them be, I found that I could live with both of them in my head without confusion. I no longer drove myself crazy trying to discover which one was "right"—an impossible situation. Then things began to move in the direction of more acceptance, in a changing, fluid way. No more collisions in the head—or at least they were more brief when they recurred in other circumstances. *This* acceptance leads to change, and explains to me my lifelong resistance to "acceptance" of the kind which keeps things "as they are." I have no use for staying in a rut, whether the rut is straight or runs in a circle. To me, that's dying—doing things to occupy me until it is time for the undertaker to pick up my body. I am out of touch with life, which moves and grows in directions unknown to me until I am within them. *Living,* I never know where I'm going next. I like this.

It seems to me to have been the history of man to go from belief (concept) to belief (concept) and each time to get stuck. The beliefs were useful (releasing) for awhile—a fresh way of looking at something, leading to change—but we accepted them *as* reality and so didn't notice when they became a handicap because we had stopped questioning. A final answer is the end, and that is death. When the feeling of dying (non-living) became too cramping for too many, it took conflict and upheaval for a new belief (new way of looking at the world and life) to come into being. This has happened in me, too. But I think that a new era can happen now, just as it is happening in me, and in those scientists who say that there are always alternate theories—probably an infinite number of them; that they select those theories which they like, which they find useful, and that "you cannot prove your theory correct by its consequences. . . . We are already very hapy if our theory does what we set out to do, and need not demand that beyond this we show that it is the only theory which achieves what we set out to do." (Karl Popper in a typed transcript of a BBC discussion with Jean-Paul Vigier.)

Warren Weaver writes that "scientists—even the greatest ones in the most advanced field of physics, such as Einstein and Bohr and Planck and Dirac—cannot agree as to whether and how science explains anything." He loves science (and also life and himself) and writes that:

> This imperfection of science I find a most attractive one; for it reflects the fact that science is not monstrous and monolithic, but is a very human enterprise, exhibiting the same lively and useful diversity which one finds in philosophy, art, music, etc. . . . We must bring science back into life as a human enterprise, an enterprise that has at its core the uncertainty, the flexibility, the subjectivity, the sweet unreasonableness, the dependence upon creativity and faith which permit it, when properly understood, to take its place as a friendly and understanding companion to all the rest of life.*

* From "The Imperfections of Science," by Warren Weaver. *American Scientist,* V. 49, No. 1, March 1961, pp. 107 & 113.

We can begin now to live without dogma and the conflicts which arise out of dogma within myself or between myself and others, and without getting stuck. We can live with the ever-changingness to which Carl Rogers finds that people are released when therapy is successful.

I and Other I are a way of looking at something in myself. I have found these concepts useful in unmuddling me, and some others have found them helpful too. But even if this way of looking at "me" appeals to many people, it does not make other ways "wrong." It only says that we like this way, that we find it useful to us at this time.

I can't remember now how many other ways I tried before arriving at this one. I only remember that whenever I had got the hang of something I thought, "There must be something simpler; there must be something more in accord with our time." This expresses my own preference for simplicity. (At this time, in this respect.)

Some things have been helpful to me but at the same time quite seriously harmful. For a couple of decades I accepted this as "inevitable"—a "fact of life" because I noticed it happening all over. But now it seems to me that anything would have been more helpful and less harmful—whether it came to me through people or through their books—if I had clearly recognized that the concepts which were presented to me were concepts—not reality. It is when I think of concepts *as* reality (which is the way they were presented to me) that I get in trouble. As concepts, I am free to choose for myself, use the ones I choose for whatever I can get out of them, and when I've used them up, move on. This is what I enjoy living with in my life and in the lives of others. Science is a product of man and in itself is not either human or inhuman: it *is* according to the humanness or lack of it in the people who use it. The *same* information on the effects of nuclear fallout is used by one scientist in a way that commits him to work toward disarmament, and by another to insist that we must develop more bombs.

My humanness is decreased when I exclude any part of *me*.

When my son graduated from Caltech I watched the men of his class marching up the central aisle to the platform, and something that I saw in all of their young faces filled me with such sadness that all of me was weeping. The tears began to flow.

What I had "seen" that made me weep was all the young men marching through the centuries from the time of the first academic processions, young men bright with hope and with accomplishment after so much effort, so much giving up of life to this—*for what?* They marched through history (in my head) as though to slaughter.

This could be interpreted as a "vision." I suppose that the word "mystic" should go along with that. But I don't like that interpretation so it is not for me. *I* prefer to see it as like my dreams which tell me what I know but haven't been noticing. When I did see this knowing that I already knew, it overwhelmed me.

I stopped my tears for the same reason that so often I stop my words—because I would be seen *in reverse.* My knowing and my feeling, that produced my tears, was both personal and impersonal, with the personal in place as a small part of the whole—not lost, much present, but not exaggerated to become the whole which it is not. But others would see in my tears my son and me alone within an alien (to me) context—within some arangement (according to each person's own thinking at the time) of the proud/sad/happy/Mother at her son's Commencement, his graduation from a particularly honored school.

To have my tears *mis*understood, to be comforted for what I am not uncomfortable about, to be "understood" through another person's understanding which is not mine, to be read *backwards*—this jangles me. It takes me out of what I am in—the source of my tears, the reality of me at the moment, the totality of myself—and because I cannot accept what others *think* that I am in which is not the truth of *me,* I am left in nowhere. A connection has been broken. I reject the other, shut him out, so that I can re-connect with the truth of *me.*

I feel deep accord with a schizophrenic client about whom Carl wrote me, saying, "I told him that being with him was like being with an Indian, because we both sat silently not saying anything, and yet I felt a good deal was going on between us. For the first time in a long time, he really laughed at this, and I am sure he got the point."

I feel that I would have really laughed too, with relief and release that someone shared my awareness, which is not acceptable to so many. They tell me that it isn't so; or that if it is, I shouldn't have it.

An artist friend said yesterday on the telephone that it seemed to her that when there is a serious family quarrel, non-verbal communication is being ignored. I told her what I have already put in this book, about the doctor who told me "I'm so glad this happened to you, not someone else!" and she laughed joyously that her knowing was shared. I had given her only the sketchiest picture of what happened, but she understood totally. With some people I had to give many more details, and some puzzle over it and never really understand, because there is only intellectual understanding. That is like knowing that a plant needs sunshine or shade, water, an appropriate soil, but never having a *feeling* for the plant as needing these things in the way that *I* need what I need. There is no responding to *this* plant, noticing and observing it, *caring* for it—not overly, excessively, even neglecting it at times or going *against* the "rules" of its behavior when this seems right. When I do this, using all of me, not only intellectual knowing, then I have the "green thumb." In Hawaii where this is also recognized, as it probably is the world over, it is called *lima ulu* which means "the growing hand."

A schizoid friend and his wife are in constant difficulties. She approaches things intellectually, while he gets tired of words, tired of the misunderstandings that come about through words, tired of the very sound of words going on and on (no silences, no pauses). Sometimes he says that he is giving them up, to live with music for awhile. His wife said to me, in complete puzzlement, "How can you communicate without words?" But her husband's words do not communicate to her easily. Even in very simple things, she has to struggle. When he referred to a jazz musician as a "skin and bones drunk" she was baffled. She tussled with it, working it out in her head, and then she said, "Oh, I see. You mean that he is thin and gets drunk. You see, I didn't understand the modifier." She teaches English, which can be a handicap. When I was working as a university press editor, I mentioned to a most lovable English professor that a five-line sentence in his manuscript didn't say anything. He read it and agreed, but then went over it phrase by phrase, clause by clause, adjective by adjective, and showed me that the sentence was absolutely correct, so therefore it should remain. It seems to me that most psychologists must do something like this with people: they don't seem to read people very well.

I don't even know what a modifier is. I think that my friend's husband probably doesn't know either, but he has a great respect for the education that he hasn't had, that his wife has. That's how mixed up we get—all of us, in one way or another. Actually we get mixed up in so many ways that I don't see any way out of it but to go by our own experience, and to limit ourselves to that. I do better when I know that if I have not had the experience, I cannot know anything about it, and can neither approve nor condemn. I can only explore and inquire, if interested. "No one may condemn another until he has walked in his moccasins for ten days." And then, of course, I won't. This sounds passive, and "passive" is a bad word in our society (another tabu). But to *experience* "passive" is to know being very much alive, and it is the source from which my right actions spring.

To think about *anything* without experience is nothing. In the early years of World War II, General Petain, Marshall of France and Chief of State, was condemned for having collaborated with the Germans when they invaded France. It did seem "bad" because of the way that it was presented to us. But when I sat alone at night with my sleeping children after Pearl Harbor, expecting invasion, and considered what I would do when it came, then Petain looked very different to me. I tried, most honestly, most searchingly, to know what I should do when invasion came. Should I take the children and run to the hills? Should I fight the Japanese when they came to my door? Should I go out to fight them? Should I make friends with them and hope through that to be able to ameliorate matters for others besides myself? I still don't know what I would have done because it never happened (to me).

When I was young, I had notions about what I would do if I found a burglar in the house. My friends and I used to talk about things like this and say what we would do. What a lot of life we waste! When I actually did meet a burglar in my home—opening the door to my bedroom, I found him directly on the other side, looking into a top bureau drawer with a flashlight—I stood there and exclaimed, *"What do you call this?"* The words sound idiotic when it is obvious what anyone calls it, but they were an apt expression of my astonishment.

All my thinking *about* meeting a burglar had left me totally unprepared for the fact itself.

When war came to Hawaii and I *experienced* it, I thought

All that I have *read* about war
All that I have *heard* about war
All that I have *thought* about war
And it isn't anything like that at all (to me)
I am completely unprepared

I had the feeling of waste—of wasted life, of wasted time when I might have been living. As I lived within war, close to others within the same war, I saw how different it was for each of us. A year later, when one evening I felt like writing about it, *my* story began, "There is no story of Pearl Harbor. It is many stories, each man's his own, made up of his position at the time, his consciousness of war about to come, whether he was alone or had a family to fear for, his own inherent sensitivity to pain, his quickness of perception." (I would now leave out the word *inherent.*)

The awareness I did not have in the early morning of December 8th when my husband came home from duty at a first-aid station, stumbling up the road in the dark. We went to his room, where the children asleep in the living room could not hear, and he told me how very bad things were, and then he said, "This is the end of everything for me." Usually I had been able to accept my husband's world even when I did not understand it. When he stayed in his room for three days at a time, as he did periodically, I brought him orange juice now and then and left him, as he wished. I knew that he was living with his agony in his way, and that neither his agony nor his way were mine: that there was nothing that I could do for him but what he wished—bring him orange juice and leave.

But on that morning after the Japanese attack, and having been awake all night while the children slept, I was so filled with *my* world that "the end of everything for me" defeated me. "For me." Within *my* world it might be the end of all of us. I was painfully afraid, and having to stop the forming fantasies of what might happen to the children. At the same time, I had a sense of grace that so far we had been spared, and a misery for those whose lives had been chopped off so senselessly. I was concerned about our friends on the other side of the island, about whom we knew nothing. "For *me*." I felt a kind of numbness as I walked away, leaving him sitting on the edge of his bed.

Yet I knew enough then about my husband's world to understand something of how this was *to him*. I even knew the terror of it because once, before we were married, I got swept into his agony, his suicidal chaos, and all the world became black to me in a most frightening, unmitigated and unmitigatable way—everything going haywire and out of control. I wanted to kill myself. I said so.

He spoke to me as though
I were a peasant
Who thought she knew
The passions of a king.

He thrust me out.

In two shakes I thought, "Whew! I'll never get into *that* again." In thrusting me out, he got out of it himself (when he became the king?) but to be in it at all was too dangerous for me.

Of course I have wondered through the years since his suicide whether there was anything I could have done to help him. Everything I read or heard of psychotherapy I thought of in relation to my husband, questioning, and always I got the same answer: "With him, it wouldn't have worked." This made me wonder whether I were white-washing my own failure. But if so, why couldn't I ever be satisfied? Why didn't I let it drop? Did I know that I was guilty, deep down in some sub-non-unconscious way? (When I mistrust me, there is no end to it, because then what I live with is mistrust.)

If I were guilty, why couldn't I feel that?—instead of living with the knowing that I had tried very hard, I had loved him very much, and I had stayed too long. I received a letter recently from a woman who stayed too long with her husband, in which she said "I have tentatively concluded (for me) that more time and more suffering than make sense are required to get on top of the mores and our personal motivations and unenlightened advice, to work through to the point where you can know yourself free of guilt for the life you can't save."

It was only last spring, nearly twenty years after my husband's suicide, that I understood something that I might have done. My relief expressed itself in a sigh (letting go). I felt better about it. Not good, but better, and better is a kind of good. And then I knew—by feeling it clearly within myself—that the "bad" that I had been feeling was not guilt but *sad.* My "better" feeling was because it wasn't solidly *inevitable* that people like my husband must suffer so much all their lives and wipe out other people along the way.

I don't *know* that I could have helped. But I *could* have been more in touch with my thoughts and feelings (acceptance of myself) and more acceptant of his distress, sharing it with him. Instead of both of us trying to shove out his distress and replace it with a happiness which was phoney, I could have let his agony *be,* and then there might have been the beginning of change. Where to everything else I had said "No, it wouldn't have worked. He wouldn't have let it," when I thought of *this* in relation to my husband I remembered the times when I *might* have got through to him. Even if I hadn't, I would have been more *me* and that would have been better for all of us.

The most beautiful times that I remember with my husband are the many hours that we spent together saying only a few words now and then, like a touch on the shoulder or a hand touching hand. Our chairs were yards apart, but we were together. At such times he was in touch with the wordless resonance, the love that was in him and in me, and let it live awhile.

But then he would be sure that I, eighteen years younger, *must* be unhappy, *must* want to be "doing something," and then I lost my awareness of all that was going on in myself and between us, and the sensing with all my senses that I had been enjoying, and I thought that I had to reassure him. He never did entirely take my word for it, and I never asked him, "Does it feel that way to you?"

My husband was exceptionally good at explaining astronomy, anthropology, medicine, English literature and history, Greek and Latin poetry, and probably some other things which I've forgotten. But in the personal sense, he explained nothing. Perhaps he was so out of touch with himself that he couldn't. There were times and ways in which I understood (or heard) him no better than he did himself, and that was because at those times I didn't hear *me* too.

In the first year that I knew him, we played Russian Bank quite often. He had been winning but then he began losing, and one evening he threw his head and arms on the table and sobbed, as though he had lost not a game of cards but everything in the world. That was what I noticed at the time. Instead of letting the thought itself be said, the spontaneous noticing, matter-of-fact, without opinion, neither tender with pain nor hard with accusation, I let myself be scared by seeing this "strong" man fall apart "over a game of cards," by seeing his big rib cage heave like a child's. It was cultural conditioning that made me view this as "wrong" or "bad"—something that "shouldn't happen" and so must be stopped. In losing my neutrality, I wanted to put him back together. I put a hand on his shoulder. I bent and kissed the back of his neck. He pulled together and smiled at me, his eyes saying thank you. And I thought that I had helped him, too.

There were these moments, brief and rare, when my husband's guard was down, when he was cracking. I had botched them. I did just what he did—tried to put "him" back together, not knowing that this was not *him*. If I had not done this, just possibly he might not have done it himself. There might have been the beginning of an opening.

All those years of reassurance were crutch-type helping, and all the times when I prevented his suicide never really saved his life, because *his* life was what he never lived.

If I had said my own thought as simply as I felt it, as it came into my head, my *first* thought that was my own direct and immediate response and really *me,* isn't it possible that he might have told me what "losing" meant to him? And if I had accepted that (not saying that it wasn't so, that there were many ways in which he won, was running ahead of others—which was true in the view of the society of which we both were a part, but not in the view of all societies) might he not have been able to speak what he never could say to anyone and so could not hear properly himself?

Suppose, that morning after the Japanese attack, I had said, "You sound as though you feel this really *is* the end for you. I hope it isn't." Those were my true thoughts, my *own* thoughts at the time, that I let get blocked by what I saw as *my* world and what I thought that his *should be.* My world at that time in fact was *not* superior: it required that he be in it with me. My world did not see that he *was* in it only more so, and blocked my own immediate and direct response to *him*. My tenderness for human suffering did not include *him*.

Thomas Szasz writes that in his opinion there is no such thing as "mental illness." He would like to get rid of the notion of mental illness because it has outlived whatever usefulness it may have, and now functions as a myth.

> As such, it is a true heir to religious myths in general, and to the belief in witchcraft in particular. The myth of mental illness encourages us to believe in its logical corollary; that social intercourse would be harmonious, satisfying, and the secure basis of a "good life" were it not for the disrupting influence of mental illness or "psychopathology." However, universal human happiness, in this form at least, is but another example of a wishful fantasy. I believe that human happiness is possible—not just for a select few, but on a scale hitherto unimaginable. But this can be achieved only if many men, not just a few, are willing and able to confront frankly, and tackle courageously, their ethical, personal, and social conflicts. . . . The phenomena now called mental illnesses [should be] looked at afresh and more simply, be removed from the category of illnesses, and be regarded as the expressions of man's struggle with the problem of *how* he should live.*

Out of my own life, I go along with that.

Recently, I struggled within myself on "*how* I should live" within an experimental situation, experimentally. That is, the situation was experimental, and I was doing my own experimenting within it. Essentially, for me, the question was "How can I fit myself into my own society which gives me trouble?" I was trying to find a way to do this. That was the wrong question. I got into a mess. All my nerves felt jangled, and running around and through each other, bent and twisted, snarled, fouled up. There was a monkey-on-a-stick feeling—angular jerkiness, and being pulled by strings. I felt pushed around as though I had been bodily pushed around—bruised

* From *Law, Liberty, and Psychiatry*, © Thomas S. Szasz, M.D. 1963, The Macmillan Company, New York, p. 16.

and battered. One morning when I awakened, there were all these messed up wires in my head—some of them broken—and my helpless, angry, tortured thought was, "They're lousing up my brain circuits!" I felt *done to,* and about three hours from a bed in a hospital for schizophrenics.

I had lost all awareness of myself as an active participant. I had lost the realization that I am always an active participant, even if I do nothing or "go along with" or submit. When I reached for this awareness, I could make only the most tenuous connection with the information in my head. I latched onto the words "I got into this *myself*" and clung to them like a life-raft, but it was several hours before the details of how I had done this came back to me, and I was released to knowing my own part in the events, and again felt free. Insistence on "I got into this *myself*" led me to reject, each time they came into my head, the accusations that others, by their actions or inactions, had "got me into this," and this cleared the way for the knowing to come through that without my own wrong action, no one could have "done" anything to me at all.

In the year before I left my husband, he had almost totally convinced me that I was "insane," unfit to be a mother. "No judge would let you have him (our son)." In fighting not to blame my husband for my hurt and confusion, because I could not bear to have my love change to hate, I arrived at knowing my own part in it, and told a doctor what was true: "I watched him dig the pit, and obligingly fell in."

But I could not see that clearly while I was under fire. Each time that I almost grasped it, my husband's words—or even the way he looked at me—threw me into a chaos of uncertainty again—the doubting of myself. No one else seemed to find *his* behavior odd, so was it all in me? I had to be away from him, away from everyone with opinions, to find out.

It gives me the creeps that children who are told they are "wrong," "no good," "worthless," cannot get away, and that our young people are required to fit into a society which does not fit them. When we require always more doctors, more psychotherapists, more jailers, more cops, who is at fault?

*A Client-centered Approach to Schizophrenia: First Approximation**

John M. Shlien

> The truth shall make ye free—but first it shall make ye miserable.
>
> Proverb

In the complex and challenging field of psychotherapy with the psychoses, client-centered therapy is a relatively new entry. The orientation now known as "client-centered" and formerly by the less accurate term "nondirective" is itself less than two decades old. Unlike most clinical or therapeutic developments, it has grown up in an academic setting, where its rate of growth and degree of influence has increased enormously in the past several years. Where it was once considered a radical view, it is now accepted as a partial technique and a general attitude by many orientations, and much of the power and originality evident twenty years ago is obscured. Its influence has spread into the guidance and counseling fields, education, religion, sociology, and social work, as well as industrial and clinical psychology; but it

* This is an abridgement. Less than half of the original paper is reprinted here. In two long sections which have been omitted, Shlien describes his view of the psychotic situation as "having an impossible life to live" and discusses self-deception and self-negation ("Complete loss of self—of all selves—is the ultimate dread faced in the acute psychotic state"). This is powerful and illuminating reading which has been omitted here only because it is less directly relevant (although certainly relevant) to the theme of *Person to Person*. The full paper appears in *Psychotherapy of the Psychoses*, edited by Arthur Burton, Basic Books, Inc., 1961.

has only recently begun to develop in the field of psychiatry. Carl Rogers, central figure in the development of this orientation, is a psychologist; his main experience, and that of his students, has been outside of medical settings. This has limited our contact with the frankly psychotic who often may need, or be seen as needing, hospitalization. Thus most of the theory and practice has been applied with patients who would be called neurotic. Rogers once held, in fact, that this form of therapy was not applicable to psychotics.

The situation is changing. Therapists have developed in depth and capacity. As more deeply disturbed clients have been seen and helped, further reaches of maladjustment have been explored. Some of these more deeply disturbed clients had been treated elsewhere by physiological or psychotherapeutic means which failed. Some were seen "inadvertently," beginning, for instance, as moderately anxious maladjusted cases which developed much more severe manifestations as defenses peeled off. Also, client-centered therapists have been called upon to demonstrate in VA and state hospitals whatever contributions they might make in work with "psychotics." (This is a word which we do not yet use with conviction or comfort. It is wrong enough to call a person with diabetes—an established disease entity—"a diabetic." That is far from being all he is. He has diabetes; but when his sugar-insulin balance is normal, so is he, in that dimension. Psychosis may simply represent a mode of *fluctuating* adjustment to a realm of experience within everyone, so that while the existence of "a psychotic *state*" is undeniable by definition, the term "*a* psychotic" is very questionable.) One more influence in this trend deserves mention. A growing number of psychiatrists have taken an active interest in client-centered therapy, usually having invented its rudiments independently, often thinking it most appropriate for their own self-directed personal therapy and wanting to learn more of its spirit and formal elaborations. One over-all mark of this trend is the appointment, a few years ago, of Dr. Rogers as joint professor of psychology and psychiatry at the University of Wisconsin.

The main outcome of this trend is that at present we are engaged in the earnest study of treatment of schizophrenia. Although we cannot speak from extensive experience, intensive experience abounds wherever there is therapeutic contact with the schizophrenic condition. It seems certain that this phase of work will bring about modifications of theory and practice, for no one can face the full individual and social impact of psychotic encounters without a "shaking of the foundations."

As Freud said, "Much is won if we succeed in transforming hysterical misery into common unhappiness." In this chapter we seek to understand the schizophrenic psychosis in more literal terms as an extreme form of an all-too-common unhappiness: self-deceit. What follows is (1) theoretical discussion of the nature of psychosis, (2) a general statement of our therapeutic principles, and (3) a case to illustrate the theory and the therapy.

Psychosis

Some Fundamental Questions: Within this discussion of psychosis, many basic questions need to be asked. The answers will be incomplete and personal, representing no "official" position because there is none, and having no claim to scientific certainty via research. Still, questions of this order must be asked: "What *is* psychosis? The psychotic experience? How can psychotherapy help? What is psychotherapy? Are there psychological laws of behavior that apply to this strange, bizarre, confusing world of the deeply disturbed person?

In this area, the words themselves are vague and insecure and our ignorance so great that we tend in desperation to assume meaning where none exists. Let us recognize that the true nature of psychosis is a mystery. (And the nature of the most prevalent convulsive treatment, electroshock, is called "a mystery within a mystery" by the most authoritative book on the subject.) One of our problems, then, is how to deal with a subject consisting of experience which at its worst is indescribable from the inside and incomprehensible from the outside, and this without using words which are themselves confounding. "Psychosis," for instance, has an authoritative, antiseptic sound, but its real sterility lies mainly in its lack of clear meaning. It simply replaces "madness"—now a literary term, and "insanity"—which represents a dated legal concept. Falling into pseudoscientific conventions of language will not help. At the present stage of knowledge the questions are well enough represented by asking simply: What does it mean to "lose one's mind"? How can a "lost mind" be recovered?

For that matter, how is the mind developed in the first place? It is our assumption throughout this chapter that "mind" develops and exists beyond brain, and following from this, the assumption of social psychological origin of much mental disturbance. It sounds simple, but there is by no means wholehearted agreement in the field on this issue. Current work on molecular structure and the chemistry of schizophrenia, for instance, challenges psychological assumptions. It is possible, of course, that biochemical methods of treatment will develop actual cures for psychotic states, thus outmoding psychotherapy. No less a therapist than Freud thought so in regard to schizophrenia. Also fashionable are experiments with drugs which induce pseudo-psychotic states, which suggest to some that if such states can be caused by chemical means, they can also be cured by chemical means, and further that the mechanism of disturbance is fundamentally biochemical. Undeniably there is always a biochemical basis for behavior of the human organism. But this does not rule out psychological influence, in either the sickening or the healing process. It is certain that anxiety can cause diarrhea. Chemical mechanism? Surely. And a virus or a laxative might cause the same apparent result, but that would not alter the fact that anxiety, a psychological state, can and does cause diarrhea (as surely as a nonchemical state may cause someone ten feet away to blush). Nor would it mean, more obviously but no more truly, that even though the eventual chemistry, mechanism, and result are the same, two different causes (laxative and anxiety) are therefore the same. Nor would it mean than an antidote for diarrhea is a specific treatment for *either* anxiety *or* a virus, or that a specific for one is of any use for the other. Similarly, a chemically induced psychotic state may not be a true psychosis even though it has the same appearance in terms of hallucinations and like effects. A chemical antidote for the pseudo-psychosis is not necessarily effective for a true psychosis, even though a chemical mechanism exists in either case. Psychological influence is not eradicated by the artificial imitation of its effects. Even though "tranquilizers" will tranquilize, so will a blow on the head, and neither one is equivalent to, or can deny the existence of, "peace of mind."

That which we call a "psychosis" is not a disease. It is a learned behavior, exaggerated to a point of no return, i.e., where control is lost and the exaggerated behavior "takes on a life of its own" temporarily. Because this exaggeration is so overwhelming, so much beyond our ordinary capacity to assimilate, it appears to us that we are no longer dealing with, for instance, ordinary suspicion, but something *quite* different—"paranoia." Then it appears that psychosis is not of the same order, not on the same continuum, as "normal" or "neurotic" behavior. But as psychotic behavior becomes more common it is seen as a form of maladjustment similar in kind to lesser degrees of maladjustment, though so much greater in quantity that it seems different in *quality* too. There is one sense, unfortunately, in which it *is* different. A boulder balanced on the edge of a precipice can be pressed ounce by measured ounce toward rolling off. Each ounce is just like the last, but when the quantity of pressure totals to the "breaking point," the quality of the *consequences* changes radically. No longer will the relief or counterpressure of one ounce recover the balance. Even if the boulder is not smashed in the fall, an enormous effort is required to restore it to its original position. It is because of this effort (which so few can make, and so many need) that it is necessary to prevent the "psychotic situation" in life. The "psychotic situation" is a precondition to the psychotic state, which may or may not follow.

Personality and Psychotherapy

A General Statement: Turning to client-centered therapy, this section will briefly review some salient features of our general position. Why "client" instead of "patient"? The negative distinction now seems somewhat picayune and, in hospital work, more embarrassing than useful; but it was intended to avoid the doctor-patient relation, with its implication of "sickness" requiring "treatment" from someone who "understands the patient better than he understands himself." A positive reason is that we really want to think of that person as a person—someone unique with dignity and capacity, worthy of our unreserved respect. If we must have a word for it, "client" seems more expressive of that. Why "client-*centered*"? This matters much more. Again, the negative purpose is only to distinguish between this and "therapist-centered," "theory-centered," "society-centered," etc. There are such things. The positive meaning is immense. It expresses the major goal of the therapist; to understand and accept the perception and feelings of the client; to share the client's view of reality rather than to impose his own.

It follows then that we rely heavily upon the *growth capacity* of the individual. The therapist is an active and significant person, but he cannot heal—he can only help to create conditions in which the natural regenerative powers take effect. As for *motive*, the drive toward self-actualization is a primary one; every human being would rather be better than worse, and strives toward the enhancement of self even though the strivings may often be thwarted and regressive. These assumptions about motive and capacity combine into a general *ethic*, which can simply be called self-determination. Client-centered therapy is founded on the conviction that man should be free, and to this end makes freedom a major means in the therapeutic experience. This is not, as some think, a reflection of political attitudes which supposedly prevail in America. It is a personal psychological conviction that the man who is most free will be most healthy. Freedom means the widest scope of choice and openness to experience, therefore the greatest probability of an adaptive response. For the individual, it seems that the urge to freedom is an urge to health, and precedes, rather than reflects, a political order.

Other Characteristics

There is in client-centered therapy an overall characteristic—an inclination toward the most literal of explanations.* There is a deliberate shunning of the arcane, the esoteric (and, some say, the erotic). Empiricism and "commonsense"

* It has always been galling to me to hear client-centered therapy, because of its techniques, referred to as passive or superficial. It is often quite the opposite, though dazzling interpretations and other intriguing esoterics are missing. An Australian colleague with whom I once discussed a particularly deep commitment I felt in a case said, "It's a life for a life." Nothing ever rang more true. This is especially so with psychotics, and I have wondered, as have many other therapists, how many times in my life I will be willing and able to make as deep an investment as seems needed.

conjecture typify the research and theory. These lean on and are expressed in terms of learning theory, perception, social psychology, interaction analysis. This characteristic leads to positions which diverge from traditional clinical opinion on many issues. It is only possible to skim over such issues here. Attitudes toward tranference, for instance, range from considering it a fiction which protects the therapist from the consequences of his real behavior and its effects, to viewing it as early stereotypic behavior which will be extinguished in the natural course of events if it is not cultivated. Symbolic analysis plays little role, partly because it is usually content-specific and covered by response to emotion. (I believe the so-called archaic symbols are significant, chiefly because they are cleaner communicants; like new words in a foreign language, they shake loose the multiple complexities of conventional meanings, and say just the elementary things each person means them to say.) The unconscious does not currently figure in client-centered literature. It is considered a reductionistic assumption about phenomena better understood in terms of attention and "levels of awareness."

An ahistorical bias has always been plain among us, and this is one of the currents which moves us to an appreciation of some of the existential writings. Diagnosis is a moot point too, though our departures are not as radical as they once were. A statement of Menninger's is one to which I can thoroughly subscribe and which seems especially appropriate in regard to the case you are about to read:

> The word "schizophrenia" becomes a damning designation. To have it once applied to a young man can be to ruin a career, despite all evidence of subsequent healthiness. A name implies a concept; and if this concept is unsound, the diagnosis can ruin treatment; the very naming it can damage the patient whom we essay to help. Nathaniel Hawthorne in *The House of Seven Gables* told us what we psychiatrists should well know:
>
> "The sick in mind . . . are rendered more darkly and hopelessly so by the manifold reflection of their disease, mirrored back from all quarters in the deportment of those about them; they are compelled to inhale the poison of their own breath, in infinite repetition."

It is not that we decry classification as such; we recognize it as a useful scientific tool. But it is dangerous when it leads to reification of terms.

Treatment of a Case

One Who "Needed Help Bad": Few things are more difficult for a therapist than presenting "a case." It was a living, breathing, sometimes gasping, sweating experience. I seriously doubt that we can really picture each other's work this way. I know we cannot do it justice. Also, there is a natural reluctance to make public what was private, even though it is disguised. (I have always admired Freud's forbearance, waiting until "Anna O." was dead before writing about her.) How to present the material? The raw data are verbatim dialogues, but that would fill a book, a sometimes tedious book. Subtleties of gestures, unspoken thoughts, analyses, would fill a shelf. They play a part. Worse still, we are conditioned to look for journalistic answers—who did it, what was wrong, where are the points of insight and the hidden keys uncovered by the detective-therapist? It doesn't happen that way. Therapy is bigger (a whole atmosphere) and smaller (moments of internal experience) than can be conveyed by the synopses we can offer.

Michael K. was twenty-eight years old when I first saw him. He had been committed by his family, and was sitting beside the examining psychiatrist at a "diagnostic staff" in a state hospital. He said he "needed help bad" and spoke of the "machine on his head." My reaction: sympathy, fascination, and challenge. I liked this boy. I wished I could help him. The timer rang indicating an allotted five minutes had passed (the hospital has 8,000 patients), and he was gently dismissed by the overworked psychiatrist. Diagnosis (no question): paranoid schizophrenia. Recommended treatment: electroshock.

Two weeks later, cardiograms and other records completed, his treatment was about to start. I had a vacancy in my schedule and asked that shock be canceled and he be assigned to psychotherapy. During sleepless nights in some of the tense nine months ahead, I often wished I had not gone to the hospital that day.

Michael served as a navy frogman during the Korean

war; tough, but scared; was "trigger-happy" on landing parties, shooting at the dark. After discharge he worked as a carpenter until auditory hallucinations became so distracting that power saws endangered him. His family put him in a private sanitarium. There he received electroshock for three weeks, insulin for seven. He once escaped by climbing over a seven-foot wall on a midwinter night. While there he was given "truth serum," and on becoming conscious saw his father standing at the foot of the bed while the white-coated attendant wrote down answers to questions. He confessed to some childhood sexual incidents which caused him great shame. Discharged with no improvement, he stayed at home. He visited a sick friend, who died the next day. Mike wondered if the cigarette he gave his sick friend had poisoned him. He went to a veterans' hospital for outpatient treatment. To the psychiatrist there he seemed "outwardly friendly and cooperative, but unwilling to talk about himself because these were 'personal problems.'" Mike had the typical attitude of his socioeconomic class toward "talkin' doctors."

His mother, who later came to see me a few times, said that "Mickey" had always been a good boy, loyal to his family, always told her the truth—"would never lie to me" (!). She first noticed "something wrong" after Korea. "He began to wolf down his food when it was so hot it would burn." During the next few years she was his confidante when he was confined to the house. She would listen to him for "as long as I could stand it, then I'd have to leave the room. I tried to tell him not to talk about those things [whorehouses, etc.]—just to forget it." She mentioned his need to have more relaxed relations with girls, nice girls, and her distaste was evident when she recalled a time she had cleaned him up with medicinal soap after a visit with other boys to a prostitute.

The father was a first-generation immigrant, a cabinetmaker, in poor health and irascible with his sons, though meek and polite with authorities. There were three sons in their twenties and one older married daughter.

Early Phase: Mike came from his locked ward to see me at my request. I told him I would visit the hospital twice a week and would be there to talk with him as part of his treatment. He plunged into a description of the "machine on his brain." He thought "on," and it went on, broadcasting his thoughts over radio and television; he thought "off" and the broadcast stopped, but his thoughts continued to repeat in his brain. "Off" caused less shame than the broadcast, but more confusion and loss of sleep. He had "nothing to hide—a clean record from Washington on down." The thoughts changed in voice—maybe an old man, a woman, or a little boy. He had been to the F.B.I. to complain. They had laughed, and he laughed as he told of it, but said it hurt him at the time. His mood shifted from friendly affable amusement at being a source of entertainment to feeling extremely angry and resentful at being exposed. For some of his thoughts were "rotten," expressed in obscenities which "shouldn't be broadcast—what if children are listening, for gosh' sake."

It frightened him to think that nothing in his past could be hidden (he had indeed been grievously exposed at many points), but he reassured himself that he had done nothing wrong—"a couple of *mistakes*, yes."

"They" had screamed at him this morning that he "has a fruity voice" and he'll "have to suck dick" to get out of here. "Telling me I'm a fairy—I'm a *man*. I been a man since I was nine." Whether there was real homosexuality or not, Mike at least expressed a defiantly false sureness of his manliness, his independence and power to cope with the world.

In the next two weeks we began to talk to one another. Mike felt relieved. The voices subsided in volume, talking instead of screaming. He asserted complete innocence—except for a few mistakes, such as blaming the Masonic temple for his problems. He struggled to face himself. "It's pretty hard for a guy to tell a doc what's really wrong with him." As he moved closer to this effort, he moved further away defensively. Now he bore no responsibility for his thoughts, much less their broadcast; everything was put into his mind. "They can even shove images right into your head. Maybe it's Hollywood. Maybe it's the F.B.I." Now his moods switched around to the theme, "Why me?" reflecting his vacillating self-concept.

"I'm nobody. Why pick on me? I'm just an ordinary guy, a worker. I don't make any trouble or had much education—I'm just an ordinary guy. (Brightly.) I'm the only one in the U.S. it's done to. It's something for me to find out for everybody. They tell me I'm a movie star. I'll go down in history. (Suddenly deflated.) I'll go down in history—yeah. I couldn't make a pimple on a movie star's ass."

I say, "Mike, sometimes you feel worthless, and sometimes great, don't you?"

"Yeah—I dunno. I—what am I? I don't know. Just nobody, I guess."

During the next weeks he began to challenge me: I was like everyone else, knew everything, would not tell him about it. He didn't want to hurt my feelings, but did not hide his anger and disgust. We were still friends.

He wondered what his "mistake" really was. For he too was "journalistically conditioned"—to find the guilty culprit, to track down the fatal flaw. Maybe it was picking up a blonde girl in his car, thinking she was a movie actress, then spending weeks trying to find her again. Then he left the city, alone for the first time, feeling isolated. Some fragments of the interviews of this period illustrate typical interaction:

Mike: I drove up past the Great Lakes Naval Base—the electronics station there, and I thought—maybe that's where it's done. It all seemed so strange. I was all shook up. I felt so, so—ill at ease—so ill at ease. I got to a little town, got a room. Then I started to cry.

Therapist: I guess you mean you were trying to get away, and it wasn't making you feel any better—you still felt strange and scared.

M.: Yeah—I felt so strange—I went—can you please tell me where to find a Lutheran preacher? They told me, and I went to church and prayed, "Jesus Christ, please help me now," and I tried—I wanted to talk to someone—to ask 'em if it was all recorded, and that I didn't do it. And I said a lot of foolish things, blamed a lot of people.

T.: Wanted someone to know. Someone to understand.

M.: Oh they *all* know. You know, Doc—everyone knows, but no one will tell me. Maybe I'll have to pull the silver out of my teeth, but that won't help—my dad caught some of it, and he has false teeth. I've got a right to live a normal life, Doc. Why don't people understand? A man makes a mistake, you don't have to murder him, drive him out of my mind. (Strikes match ferociously.)

T.: (Could respond to the admission of error and resentment of punishment, but impressed by match.) Makes you *mad.*

M.: Hell, yes, (Glares.) Then I heard these two guys at the plant say, "We can't turn 'im off yet, we got a sum of money invested in 'im."* Now, why don't they tell me what it's all about?

T.: Why won't *some*body level with you?

M.: That's right. Why won't—I've had a hell of a time, Doc. I'd hate to see anybody go through this. Seemed like if I tried to explain this to somebody, they'd think I was nuts. Why won't someone believe me? They laughed, the F.B.I. I guess they got a charge out of it.

T.: Makes you feel there's nobody on your side, nobody who sees it your way.

M.: They don't seem to. I been to a lot of places. Doc Millman, I told him, and he said, "I don't see no wires on you, Mike." Yeah, don't see 'em—well, I don't know how it's done, it's electronic or what. Anyway, I know I'm personally not to blame, but they can't help me.

T.: How about me, Mike?

M.: I don't know about you. (Pause.) If you could go through it, maybe a week, maybe a few days—but I'd hate to see anybody else go through this. But if you could, just for a couple of days—they can illusify your mind, like I seen (movie actress) and she called me

* Probably distortion: the men are in a woodworking machine plant, and it takes only a little twist to misunderstand *them* (the machines) as *him.*

over, waved, like this, and she was so real, you could reach out, shake hands, dance, do anything you want....

T.: So real, so vivid—it's an amazing and marvelous thing. I guess you're telling me maybe I ought to experience it too, but—(Mike looks startled)—what's the matter?

M.: They just said, "What a son of a bitch you turned out to be." Screaming it at me. I never done anything. I never killed, raped, or crippled. People put thoughts in my mind. I'm all confused. Everybody must know I'm not a bad guy. (Begins to sob.)

T.: Hurts so much.

M.: If somebody was in my shoes, they couldn't tell nobody. I told my fiancee. She says, "See a doctor." I told my minister. He says, "Mike, better see a doctor." What's the use? I ain't got a chance. Not a chance in hell.

T.: I guess you're feeling, "I've tried, and nobody understands. Nobody shares with me—I'm just all alone with it."

M.: All alone, Doc, that's it. What's the use? (Pause.) Can I bum one of your cigarettes?

At the next session, Mike said he had received shock treatment that morning (to my dismay!) and had "forgotten a lot of things." He was meek, submissive, watchful for cues as to what his behavior should be, anxious to please, and showed temporary amnesia for recent events. The administration of shock was by accidental order, and was discontinued, but it was a breach of trust which affected our relation in ways I can never appraise.

Middle Phase: Mike was very morose. He thought his behavior had undermined his family's confidence in him. He felt very helpless now—caged, imprisoned, victimized—and said he got himself into this jam by blaming others (the Masonic temple, etc.). He remembered being given "truth serum," strapped in bed and struggling not to tell the answers to questions put to him. It was on this occasion that his father heard him confess his childhood experience of fellatio, for one thing. When he was taken home from the sanitarium, he thought he ought to tell his mother about it, but she was "not too pleased." At about this time Mike began to develop and/or hint at long-standing feelings toward psychiatrists as the punishers, the probers, the exposers, the unhelpful villains in his experience.

Some of this developed in the next interview. Mike was talking about his observations of a woman patient at a dance.

M.: I was trying to figure it out. Everything I thought, I saw her lips repeating. She was in the same sort of shape I'm in. She repeated every word that ran through my head. Then I thought—maybe it was the patients who were mentally ill, it might be good to use this machine on those people. It might help them. Now, it might be the psychiatrists who are doing this. Who else would want to know the function of the mind? I don't. It's a wearying thing to me. I'm pretty tired of it.

T.: Maybe this would help some other people, but it's not helping me—I'm tired of it.

M.: Yeah. Well, I studied it, up in my mind, I thought about it. I was given truth serum, and I wondered how the inventor did it, and what I worried about was *whether the inventor is responsible.*

T.: I don't understand that, Mike. What do you mean?

M.: I read books on it, and studied on it. And maybe when I told about this during truth serum, maybe that's when they heard of it. I don't know if I had anything to do with it. I can't figure it out. It's too impossible. You can't see anything. All you can do is hear—and you can see hallucinations.

(He went on to speculate that he invented it, but they [the psychiatrists] perfected it, wanting to test his strength.)

M.: Let 'em experiment with somebody else. The mind is a delicate thing. I'd hate to see anybody else go through this shit. It's not funny. I know I didn't invent this thing, and if I did, I'd like to burn it. It's like taking your fucking life away from you. I don't go around pulling switches and throwing the juice in people. (Probably a reference to shock therapy.) Now the guy at the foot of the bed—he wrote everything down, and maybe they got the idea from me. Whoever invented this fucking thing oughta be shot—then I'm thinking, oh oh, maybe I'm partly to blame for this.

T.: Seems as if somebody else ought to be punished, but then you think, "God, maybe I thought of this, I wish I hadn't."

M.: That's right. I think I kind of invented it but the psychiatrists perfected it.

As Mike began to assume some responsibility (symbolically), he defensively moved further away from ultimate blame and further into self-deception. He finished the interview in a rage at being deceived, the butt of jokes, dishonesty, experimentation.

In the ensuing week, he told the ward physician to "please cut my brains out before I kill somebody." This referred to me, for I had become the enemy of enemies. Nor was this simply transference. I *do* represent hospital, professional people who treat and mistreat, I *do* refuse to "tell him about it" (I couldn't). I do frustrate, misunderstand, and disappoint. Mike became violent in the ward, for which he was sent to "security" or given sedation. Often he refused to see me. When he would come, his "oceanic rage" was more than I could bear with equanimity—I became afraid, and often wished I did not have to see him. During this period, attendants sometimes waited outside the closed door.

Looking back, the most interesting thing is the advice given at this time. Rogers' statement that "It is more therapeutic for him openly to *be* that fear, even expressing it to the client, than to present a calm and unruffled front" was probably written with me in mind. I regret that I did not test his hypothesis. Rather, I took the "lion-tamer's" advice of the ward physician who said, "Never let them know you're afraid. It frightens them." The clinical director, a wise and experienced therapist, felt that there were really safeguards in Mike, and that I need not be afraid, but if I was—"You have to live with it while it lasts." So I presented a calm front while I worked through my fears and Mike worked out his rage. The end of this period (which lasted three months) was marked by my suggesting that we walk outside on the grounds for the hour. We trusted each other to do this, and the most significant moments of therapy took place thereafter while sitting on a bench or on the grass.

Later Phase: Mike asked for and obtained a grounds pass—freedom to leave the ward. He was given permission for home visits. His family said he was "fine." Does that mean he had solved the riddle of the influencing machine? Or that it was gone? Neither, completely. He was much less affected by it. For one thing, he recognized *some* of the images as recollections of his own past experience. This was an advance over acknowledging the possible invention of the machine—it acknowledged the content of his experience. Second, he began to deny, to his mother (and less to me) that he heard any voices. He took the view (a realistic one) that if he told us, it would delay his release from the hospital and that *If he did not tell us, we would not know.* His privacy was restored! He might say, with a grin, "I don't want to talk about it," or soberly, "I can live with it," or just "No, not any more." If his privacy was restored by human contact, what sort of contact was there in those hours on the lawn? Sometimes silence. Sometimes talk of the same sort already reported. Sometimes he began to cry softly, saying, "They talk about needing love and affection. I know what *that* means. The only good thing I ever had (his engagement to a girl) taken away from me, broken up." He blew his nose, dropped his handkerchief, and as he picked it up, glanced at me. He saw tears in my eyes. He offered me the handkerchief, then drew it back because he knew he had just wiped his nose on it and could feel the wetness on his hand. We both knew this, each knew the other knew it; we both understood the feel and the meaning of the handkerchief (the stickiness and texture, the sympathy of the offering and the embarrassment of the withdrawal) and we acknowledged each other and the interplay of each one's significance to the other. It is not the tears, but the exquisite awareness of dual experience that restores consciousness of self. A self *being,* the self-concept can change.

Six months earlier, there had been an interchange like this:

M.: Can I bum one, Doc?

T.: Does it make you feel like a bum, Mike, to take my cigarettes?

M.: Yeah. I—it's a way I feel. I hate to have to ask for anything. At home, when I'm not working, I hate to see Mom put the food on the table.

T.: Hurts your pride?

M.: I hate to feel like a bum. I guess I am a bum.

Having a ground pass, Mike could buy his own cigarettes (but not carry matches), and he offered me one, or bought me a Coke, as often as I did him. These were important ways in which he restored his feeling of equality and self-respect. More often, he did not want to talk about himself, was sometimes surly, wanted to discontinue therapy, insisted he was ready to go home. No one felt that he was "cured," but he made it look good enough on the outside and was so much better on the inside that the hospital was not helping any longer. His parents took him home. He told them that legally they could, because they were the ones who "signed him in." At one time this was "impossible" to him; his parents could *only* love and want him, *never* would commit him.

Where is Mike now? Happy ending? No—this is really a fragment, a fragment of therapy and of his life. Mike does not come to see me. Where he lives, "A person has to be goofy to go to a psychiatrist." After four months at home, he is working again at his carpenter's trade, trying to make a life for himself. He said on the telephone that he is "pretty *good.*" His mother said, "*pretty* good." Shall the curtain of privacy be drawn against the inquiry of science? I think so. My mind goes back to a scene on the lawn of the hospital. Mike said, "I went to church yesterday, Doc, and I said a prayer that I could go home and this would never happen again. I said a little prayer for you, too—that you could help me and always be well yourself." I was moved, and said, "Thank you." Right now, in my way, Mike, I say a little prayer for you. For all of us.

IX

Yes.

X

> Life can only be understood backwards; but it must be lived forwards.
>
> Soren Kierkegaard

Let's use the old term "madness" for awhile. It doesn't mean anything but what I or anyone else sees as madness at this time, and at this time this is the way that I see it.

There are two madnesses, only one of which is generally recognized. That is the one that we call "abnormal" or "out of touch with reality." The other one seems to me to be out of touch with reality too, although it is called "normal" because a majority of people are that way in our society. In *this* sense, it is a reality—like the reality in many parts of the world that most people go to bed hungry every night. This is not a reality which most of us would recommend adjusting to although where it exists it is "normal."

There are two normals too, only one of which we usually recognize. That is the statistical normal which is nobody, like the average Harvard man who has 3.5 children. The other normal is unknown because it is *me* if I had not been interfered with, what I might have become, my *own* norm—a constantly changing potentiality becoming actuality. This happens to nobody too—but it *could* happen. Why shouldn't we increasingly make it possible for this to happen?

For years I tried to make myself the first normal—like most other people. That's how *this* normal persists—by people making themselves be what other people are. Everyone trying to make himself like somebody else. The more that I did that, the more I wiped out the second normal which should come first, the one which is an ever-changing *me.*

To hold myself to *any* pattern is to cripple me as surely as it crippled Chinese girl-babies to bind their feet—a practice which the "civilized" world deplored. *My* feet were left free to grow in their own way, with confidence that then they would serve me well for walking and to stand on. Shoes were carefully chosen to fit, but not bind, my feet, with always a little room to grow in. There was not the same confidence in my ability to grow as a person, and standing on my own feet—or moving on my own—in that sense was largely disapproved of. It was only by living a good deal in privacy that I managed to be me at all.

The recognized "abnormal" is a very lonely one, because no matter how many other people are in it, I live in so much isolation. I am cut off from other people because they reject me, and my response to them has been chopped off too—perhaps in the way it happens when I am overtired and can't even enjoy doing what I like to do. It evokes no response in me. The other madness, which is called "normal," is not lonely—or rather my loneliness has got buried so that I do not notice it so much. I can "forget it."

> . . . They may remember
> The vision they have had, but they cease to regret it,
> Maintain themselves by the common routine,
> Learn to avoid excessive expectation,
> Become tolerant of themselves and others,
> Giving and taking, in the usual actions
> What there is to give and take. They do not repine;
> Are contented with the morning that separates
> And with the evening that brings together
> For casual talk before the fire
> Two people who know they do not understand each other,
> Breeding children whom they do not understand
> And who will never understand them . . .
>
> And the other life will be only like a book
> You have read once, and lost . . .*

But the loneliness is still there. When I have substituted a superficial togetherness for the basic being-together, I have

* From *The Cocktail Party*, copyright 1950 by T. S. Eliot. Reprinted by permission of Harcourt, Brace & World, Inc.

built a superstructure with no structure underneath it. That is fantasy, even though the fantasy may become a second-order reality by many people engaging in the same fantasy. I am linked with many other people through my surface, or mask, or disguise—by approved actions, the way that I eat and speak, the faces that I make while doing this, and by "being with people" according to a formula and playing a role. This is phoney, or veneer, because it is put on, but as long as I keep my mind focussed on it, I have the illusion of being in touch with other people and myself by interacting with *their* phoney surfaces. Even if I have done this for so long that the surface seems real to me it is more like apples bobbing in a tub of water, skin touching skin, with no awareness of the flesh or core. We are all apples and that's enough, isn't it? What more do you want? Not satisfied with being an apple? Just do what the other apples are doing and you'll be all right.

Not to do that is "bad." (A threat to *my* veneer, which might crack and split if exposed to the weather of humanness?)

To me, it's mad to do it.

A man told me when I was very sick, "I have never known anyone so hell-bent on self-expression as you are!" He sounded both friendly and exasperated. I didn't feel capable at that time of what I thought of as "self-expression" but I felt guilty just the same, with the piled-up accusations of the past. I had tried very hard to be "normal" but hadn't been successful enough to escape criticism. At the same time that I felt guilty, a little growingness in me protested, saying "Well . . . I guess everyone is doing that really—expressing himself in one way or another." I was very confused, and gropingly trying to understand what I do now.

"Yes . . ." said my friend. "I guess you could say that a person is expressing himself when he makes a slightly different stew. But *you*—when everyone else is happily going over the hill *this* way, *you* have to go over it *that* way." He gestured with his arms, showing the two ways over the hill. I realized then that he meant that I was trying to find my own way out of being sick, instead of "accepting it" and "making the best of it" as people said I should.

I "saw" a hill—that is, I visualized it. . . . Habits are certainly difficult to break. *It* visualized itself, spontaneously, as a dream does. I had nothing to do with it except that it happened in me. . . . There was a picture in my mind of an endless train of passenger cars full of laughing people going over the hill on one side, while I was wearily plodding over it on the other side, alone. (When I am confused, I am always weary—worn out by conflict.) I felt "wrong" for not going the way that everyone else did. But my "vision" told me clearly, "They're not happy. They're just pretending to be happy. They've done it for so long they've fooled themselves." Laughter can be a cover for unhappiness, particularly when *signs* of happiness are acceptable and those of unhappiness are tabu. Isn't *pretending* being out of touch with reality too?

When I am surrounded by pretending people, I sometimes feel so swamped by meaningless two-dimensional cardboard characters that I feel I may be on my way to the madness that is recognized as madness. I think that this may be the way that some of it comes about. Nothing *has* to come about in one way only, and the discovery of one way eliminates others only because then they are not explored. That doesn't put them out of existence—just out of mind, like all the other possible approaches to bodily illness which the A.M.A. will not admit. I think that when we have found one way, we should use it tentatively, as the best that we have latched onto at this time, and at the same time should go on exploring other ways—with the same tentativeness.

It seems to me that the madness that is called insanity may sometimes be a reaction produced by the madness (as I see it) that is called sanity, or "realism." A patient in a mental hospital told a therapist, "You want me to come into your world, but I lived there for twenty-three years and I don't like it." The patient was a very mixed-up person, but I don't think that he was mixed up about this.

John Shlien has presented the tenderness of a "schizophrenic" person, his humanness. Van Dusen writes, "The more mentally ill they are, the more they are caught in egocentricity, selfishness and uselessness." Do they disagree?

In a letter, John Shlien says "As for 'schizophrenic person' —there is no such thing. A person scared out of his wits is no person at all." I was with some people labeled "schizophrenic" in a mental hospital recently. One of them was a woman in her middle years, of outward composure and grace. She was wearing a gingham dress—blue and white checks—with some sort of kerchief around her shoulders, but she sat as though wearing velvet and a cashmere shawl—as though these things belonged to her as a matter of course. She answered all questions, conveying complete confidence in her answers. But they went like this:

"Do you know the date?"

"Yes, I've studied that. It's April 6th."

"Do you hear voices?"

"Yes, don't you?"

"No, I don't hear them. Are the voices male or female?"

"That depends on the wiring in the building."

Her face was delicate, her eyes tender, with a hint of pain. When she was asked, "Where do you live?" she did not say "here" or "in a hospital," or even "California" where she was, but referred to a farm in Illinois where she was "so lonely." That loneliness *is* where she lives.

This woman's tenderness was very clear—but she never showed any response to any of us around her. She was "caught in egocentricity, selfishness, and uselessness."

Another woman, thirty-one, came in and walked to the window, looked out, then back, and there was something haunting or haunted in her eyes. She seemed to have no awareness of any of us, either then or when she suddenly became quick and like a child rushed to the light switch near the door and flicked it on and off. From there she went skippingly to another door and opened it, and flipped the light switch inside it on and off. Then to another door, where she did the same. Back and forth, from one to another, without end. When the psychiatrist called to her to "come and sit down" she obediently did—but in just a few seconds she hopped up and went from light switch to light switch again, always flipping them twice. When she did it only once, in a moment she turned and did it again. When the psychiatrist called to her to leave, she came, and as she went through the door she said to him, making it a statement with a question in it, "I'm not a bad girl.?" When I remember that thought in myself as a child, I know her terrible unhappiness.

She was "caught in egocentricity, selfishness, and uselessness" but as Van Dusen also says, "These people are *painfully* trapped. . . ." and "He found [my description of his schizophrenic world] accurate except that his world fell in pieces and my description was coherent. . . . There were deadly places of lostness [in his]."

Van Dusen writes in a letter, "I no longer know what the technical definition of schizophrenia is. Certainly it includes a wide variety of conditions of varying severity. Some could be genetic and/or metabolic. The key for me is that the schizophrenic has found some of his subjective inner world in the outer world—coloring it—distorting it—concealing some aspects of reality from his vision. This may or may not lead to social difficulty. It can extend from mild degrees on up to the whole of the outer world becoming diseased and disturbed, flooded with the unrecognized inner."

A touch of this inner is recognized by a young woman when she writes: "The main indication to me of how and where I'm goofed up is how and where I try to goof up others. One thing I've learned. I have very little self-confidence and so operate on the defensive and resent suggestions and even help, seeing it as criticism and/or competition. I think I've had this brilliant insight before—I'll probably keep having it until I can learn it in myself as well as in my head."

Carl Rogers writes in a letter:

> What does the word schizophrenic mean to me? To the extent that it has any definite meaning at all, it means that here is a person who is highly sensitive to his own inner experience and also to interpersonal relationships with others, who has been so defeated and traumatized in endeavoring to make use of his sensitivities that he has retreated both from his own experiencing and from any real contact with others. In addition to this fairly basic understanding, the word schizophrenic also means to me that anyone whose behavior is deviant and who cannot easily be understood is tossed in this all encompassing basket.
>
> I try to remain open-minded to the possibility that there may be specific genetic, chemical, or nutritional factors which may bear some causative relationship to the behavior that is called schizophrenic. But thus far I have seen no evidence that would convince me.

In a letter, Gendlin says:

> I believe any definition of *schizophrenic* offered today must begin by stating that as yet we know very little. It must also be said that a very large variety of thoroughly different people are currently being labeled "schizophrenic."
>
> I believe that schizophrenia is *the absence of* (or great narrowing of) that felt experiential interaction process which we are as people and which we feel as our concrete selves. When an individual is stuck in a hurtful relationship situation so that he can neither leave it nor feel and relate himself sufficiently to it, then he becomes deadened and empty inside, reports that he is "not himself." The ongoing feeling process on which we all rely inside becomes muddy, swampy, unreliable, or stopped, and feeling contents become stuck, frozen, unowned, seemingly alien, static places, rather than fluidly functioning feeling.
>
> I deny that psychosis is psychotic contents that are in all of us and can "erupt." I believe instead that such "psychotic contents" are really a certain *manner* of being alive. Recent LSD and other drug experiments have strongly confirmed this view. It seems that whether an individual under the influence of the drug comes up with horrible experiences or with wonderful experiences depends not on him (as a self-enclosed piece of well or poorly working machinery) but on the relationship situation in which he is (and feels himself to be) when he takes the drug. If he takes it in a friendly situation close to someone he trusts, he has a wonderful experience. Early psychological experiments which attempted to test LSD under carefully pure circumstances (the subject alone in an otherwise empty room!) produced something very like schizophrenia.
>
> This definition of schizophrenia as an *absence* of a certain type of felt interaction (an absence rather than an illness or a content) has important consequences. It seems that we think of ourselves not so much as work-

ing with something ill in the person, but rather as providing the necessary personal relationship in the context of which someone can again come alive as a person.

John Shlien says, "As for 'schizophrenia,' how we love to think that we know what's wrong with those others."

What about ourselves?

When my husband was twenty years old he wrote a sonnet:

Winds of the sea caress my eyes and hair
The currents of the ocean swell and strain
Into my arms, and I have left no stain
Upon the laughing sea and sunbright air
The tide-race is my friend, and I may share
The stars, which suffer me without disdain,
I greet the sunlight and the summer rain
And while I love them, they are none less fair.
But when my hands reach out to other hands
My touch is venom, and the gifts I bring
Are bitterness, importunate demands,
Heart heaviness, mistrust, and questioning.
Oh, I am weary of the strife and pain.
I shall go back and love the sea again.

I did not know him then, and anyway I was two years old. Twenty years later, the way he felt about the sea was still the same, and the way he felt about people was the same, but the way that he lived with the sea was not at all the way that he lived with people.

How many of us can accept people as we can accept the sea? There are of course people who are foolish enough to want to change the sea, who want to change the weather, too, instead of living in accord with it. But why are so many of us so hell-bent on the madness of changing other people to suit ourselves, and forcing them to do what we want them to? Perhaps all the sensitive and human people get scrapped and sent off to mental hospitals? Perhaps "mental illness" is a sign of *health*—of a person who refuses to knuckle under to a distorting and inhuman society? Who is healthier, the German youth who had no conflicts about shoveling people into ovens, or the one who had a "nervous breakdown" because of his conflict about being pushed into doing something he didn't approve of?

I want neither of those madnesses. Offered a choice, I say *No* to both. When I say no to both, I have to find my own alternative.

When I am going someone else's way I require that others go along with me. When I am going someone else's way, I have let myself be sold on this—it doesn't come from me. Coming from *outside* as it does, I have an image of "what is good," and then it seems to me that all others should follow this image too, and do "what is good" "is right" "like me." At a deep level I am afraid, but it is not the fear which has made me follow the image, but having images that has separated me from myself and so I am afraid: In my fear, then I need others around me to do the same things that I do, to reassure me. When I am going my own way in freedom, I wish most for everyone else to go *their* way in freedom. When all of us are moving in freedom, there are no collisions. This freedom *looks* chaotic to me only if I have a picture in my head of "how things should be" which doesn't match with what is happening.

When I went to Hawaii in 1934 and the traffic in Honolulu looked like madness—totally disorganized—this was because it was so different from the order that I was used to. I was like the little girl who had had a crooked body all her life and when a doctor straightened it she turned to her mother and said, "He made me crooked!" The Honolulu traffic looked "crooked" to me. But as I walked around doing errands on many different days, still failing to find any order in the driving, gradually I became aware that there were no accidents—not even near-collisions—and nobody got angry at anyone else. Then I became aware that the drivers of the cars—especially the "uneducated" Hawaiians, Orientals and Filipinos—were doing what I did: noticing. There were some rules—like driving on the right-hand side of the road and stopping for pedestrians, but the drivers went clear over on the left-hand side any time that was appropriate either to moving on their way or to getting out of someone else's, and cars stopped as soon as a pedestrian stepped from the sidewalk to the street, even if this happened in the middle of the block. The driving was very much person-to-person, with recognition on both sides. Usually there was just an instant of this recognition between driver and driver or pedestrian and driver, but there *was* this moment of awareness of each other, of human "speaking" to human, and although I didn't know anyone I felt surrounded by friends. When I had got the hang of it, I went right into Honolulu and drove the way that everyone else did. I have never had so much fun driving in my life. It was even more fun than driving in the Southwest where people were scarce and noticed each other when they passed on the road. In Honolulu there was no lack of people and cars, and still there was this noticing.

In Hawaii then, when I was thirty-two years old, there were many ways in which I arrived at my *own* normal, which I could know by the feeling of ease and happiness, of having left an alien world and come home for the first time in my life. I had felt that way before with a *place*—the land, the country—as in the Southwest, but never with so many *people.*

It seems to me that the present "chaos" in psychotherapy has something in common with the "chaos" that I saw in Honolulu traffic. I have read several books in which many or several psychotherapists were included, sometimes presenting their cases separately, sometimes discussing cases or psychotherapy together. There is so much difference and differing among them that when I first read them it looked like madness: Doesn't *anyone* know what he's doing? (If they knew what they were doing, they would all be doing it the same way—going by rules which I could recognize.) Gradually they sorted themselves out in my mind into two groups: those who are following some Authority by copying him, and those who are struggling to find their *own* way, which seems to have the common base (or direction) of spontaneity, of responsiveness, of being in touch with myself and what is going on in me (the therapist) as well as what is going on in the person who has come for help. Should we live with people in any other way?

These individual strugglers, as far as I know them, have a good deal of acceptance of and respect for each other even when they disagree, and although their *apparent* methods are so different that they seem not to be the same thing. There is "client-centered therapy" and there is "communication" therapy and there is therapy through swimming instruction and there are therapists who fall asleep repeatedly during therapy sessions, as part of the therapy, and when they wake up report their dreams to the person who has come for help. There is therapy by "guided daydreams," and there is "transactional analysis" which makes clear to people the games that they are playing with each other and helps them to give up these games. Essentially, all of these methods achieve, when successful, a switch from dishonesty and competition to honesty and cooperation.

All of them attempt to relieve us of what is binding us. But when a therapist (or anyone) tries to copy what is spontaneous, immediately this is defeated. When I *try to copy,* spontaneity is lost. I *reverse* what I have seen and liked and want to follow. With my intellect I have picked up a picture which I am trying to reproduce. I hold myself to this picture of what I think I "should" do or be. The more that I do this, the more inflexible I become, and spontaneity cannot happen no matter how much I wish it to. What I have seen in the therapist (or other person) whom I am "copying" is the result of an effort toward spontaneity and free response. I am copying the *result.* The more that I hold myself to it, the farther I am from my aim. (My aim is not a goal, but my direction.) The happenings which preceded the result are missing. What should come *second* has been placed *first.* Then, what should come first has become impossible.

It is very easy for me to read myself incorrectly afterward even when I have moved in the right way. I mistake what happened for what I did myself. It happened by itself through me. But later, I take credit for myself, as for something that "I" have done. A young therapist described a therapy session to me in this way: "This fellow—a young guy, nineteen—was laughing at himself and his friends for sitting on the beach day after day laughing at themselves for the way they were 'all fucked up'—mixed up in homosexuality, doing crazy things that got them nowhere. They called themselves names and ridiculed themselves, and he went on doing this with me. I couldn't get him to *feel* anything. He just kept on laughing at what dopes they all were, including him. So then, I started laughing too, laughing with him at his being such a dope—and then he got mad, and began to say how he really felt about it. Then he knew that what he was really feeling wasn't funny." After that, they began to move in therapy.

I asked the therapist, "Was that something that you figured out and then did, or did it happen and afterward you saw the sense of it and why it was successful?" He looked a bit disconcerted and unhappy for a moment—the way that I have often felt when I have realized that what I have taken credit for happened through me but was not done *by* me (except in the objective sense that it could be observed as having been done by me). Then he said that it had happened to his surprise, and afterward he figured out why it had worked the way it did. The way that he *told* it sounded as though he had figured it out *first*.

I do this too, in part because this is our habit, but also because the other way around is acceptable to very few people. The Age of Reason insists that we figure things out *first*, then do them. Any other way may seem unrealistic. But Tauber and Green say:

> In this connection there are many situations in ordinary experience which demonstrate that much can be reacted to more effectively if consciousness does not prematurely share in the perceptual experience. There are many illustrations, for example, of man's capacity to register perception accurately in space and time categories provided that conscious cognitive processes are postponed. With this knowledge of man's superior judgment when using his precognitive capacities for certain tasks, the army trains its artillery observers to utilize their capacities to the utmost. The observers must always call the position of a shot as quickly as possible—there must be "zero delay" between noting the fall and shouting out the location. All beginners wish to estimate with the aid of rational judgment, but experience has shown that there is unquestioned superiority of performance when rational estimation is suspended. The first flashing quick guess turns out to be the best guess. *

I remember having read in the verbatim reports of the Utah Conferences on Creativity, sponsored by the National

* From *Prelogical Experience*, by Edward S. Tauber and Maurice R. Green, Basic Books, Inc., New York, 1959, fn. p. 71.

Science Foundation, that an air force testing officer reported that in their search for creative men they found that "those who get the answer without knowing how they got it are the ones we want."

The usual insistence on figuring things out first results in many things not being done because we haven't got them altogether figured out first. In an education seminar a psychologist asked, "How is it that we know so much and don't *do* anything?" A seventeen-year-old told me "talk, talk, talk, and nobody *does* anything. Sometimes I think I'll go really crazy."

In reaction against this, many people—young people especially but it is not limited to them—*try* to do something "spontaneously" and it flops. A young girl throws all sorts of food into a bowl and mixes them up and sticks it in the oven, and it comes out pretty awful. A young man wages his personal war on "alienation" by inviting other people to live with him and then it doesn't work out the way he thought it would. Neither of these two young people were in touch with their own experience, of food or people. Their *experience* of food or people would have shown them a better way of putting them together, and also in that case it would have been done more tentatively, a little at a time, observing what happens and making corrections. The goal is non-specific, and in that sense there is no recipe. But all the information from the past is available in the present. Even if I haven't cooked before, still, I have seen other people cook, and I have eaten.

When I applied for a job as a waitress in Arizona when I was twenty, I was asked if I had any experience. I said yes (to my surprise and delight—I desperately needed a job as I had arrived there with only 50¢ after I had tipped the porter on the train), that I had waited in some of the best restaurants in New York. I didn't mention that I had done my "waiting" sitting down. With that information from the customer's side, and keeping my eye on what other waitresses were doing in *this* restaurant, and noticing the responses of the people whom I served, I started out as a fairly good waitress and became a better one. I worked as a waitress for less than two months, but before I quit (to take a job as society editor) the best hotel in town offered me a job. As I have put it here, I am telling it backwards, for I went ahead and did it, and figured out later what had happened. My "smartness" I can take no credit for. I cannot feel proud, just happy, when things work out especially well. I like the feeling of happiness better—a word that springs from another word, "happenings."

When I lived in this way so much when I was young, the world seemed full of happenings. I had no experience of being a society editor, either, but newspapers had been around me all my life, and I had been writing ever since I learned how, and I had been talking, which is a kind of writing. I left school on my sixteenth birthday, before I had got trained out of the idea. When I went to receptions for famous people, to report on the reception, I went as a guest (the feeling in myself was that of a guest) and I had been a guest who talked with interesting people and learned about them. Instead of talking about them as I had after other parties, I wrote about them. In this case too, analysis comes *after* the experience. I didn't figure out how *to* do it first. At that time, this was the way I lived.

Perhaps a step by step illustration will help to convey my preferred approach to life and living, by moving into the largely unknown and picking up further information—guides and clues—along the way. When I was 22 and working at "public relations" in New York, at five o'clock one day my boss told me that there was a dress rehearsal of the circus that evening in Madison Square Garden, beginning at eight o'clock, and that no one was permitted to see it. He asked me if I thought I could get him in. Why I said yes, I don't know. I have never looked into that. What comes to mind is that (1) I was used to carrying out his assignments whether I thought I could do them or not, and (2) this was something I had never done before. My boss asked if I could get his friend in, too. It seemed to me that two would be no bigger problem, so I said yes again. *Then* I became a little afraid of failure, but I didn't want to back out.

I had been to the circus with a ticket in my hand, and that was all that I knew about it. Without a ticket, and with no audience permitted, the front door didn't seem a good way to get in, so I went to the back door. It was an ordinary door, of the usual size, at the top of three or four steps. On the top step stood a large doorman. Various people came along, skipped up the steps, bobbed their heads to the doorman, and went in. I could do that, too—but I didn't *look* like those people. There was a kind of dippiness in their clothes. So I went to my apartment and looked through what I had. There was a long-skirted patchwork print dress with yellow wool edging that I had made for the fun of it and wore only within my apartment. I didn't have a dippy hat, but I had one with a feather in it that looked dippy when put on backwards.

I went back to Madison Square Garden wearing the dress, and the hat on backwards, about half an hour before the rehearsal was to begin, and skipped up the steps, bobbed to the doorman, and went in. Until I was inside—which seemed to happen very quickly, so that I felt totally unprepared for it—I hadn't thought about what I would do when I got there. I had never before seen anything like what I was in—wide racks of shelves going clear to the ceiling, stacked like library stacks with aisles between. On these shelves were all the fantastic gear of the circus—like the horse body that the clown puts on and capers around in. I felt so bewildered that I couldn't have felt more so. I didn't know where to go. There weren't many people and they were all going here or there busily. What I must not do was stand there, so I kept moving, and went down one of the aisles. I must have had a questing look in my eye, because a man stopped and asked me—as pal to pal—"Who are you looking for?"

I said the only name I knew, "Mr. Ringling."

"He's in the ring," said the man and went on his way. I was floored, because which way was the ring? But then I realized that when he said "the ring" he had moved one shoulder a little as though pointing to it, so I went in that direction, found a door, opened it, went through—and there was the whole big Madison Square Garden that I knew, except that all the seats were empty and there was no one in the ring except three men standing in the middle talking, so I walked up to them. One of them asked me who I wanted, and I said "Mr. Ringling." Another man turned to me then, so I assumed that he was Mr. Ringling. He asked me what I wanted, and I told him. He said, "How did *you* get in here?" I told him I had come in through the back door. He said, "Anyone who can get past *that* doorman can have all the friends he wants in here!" and he bellowed for a man to come and open the front door. I went with him, and only then did I begin to wonder if my boss had been playing a trick on me, or if perhaps he thought I couldn't possibly do it and wasn't even waiting at the door. But the big doors opened, and there was my boss, and his friend. We sat in a box and were the total audience.

When I don't know my way, that is the only way that I can find it—by using whatever information I have, by moving tentatively, step by step, by *noticing* and responding to whatever happens in the moment. I think this must have been the way of successful pioneers.

I like this way of living. It is the way that this book has been done: I couldn't analyze or outline it before it happened. I think that its greatest faults have come in when I have tried to guide it. It seems to me that this is very much like Carl Rogers' preferred way of being in psychotherapy: move into the situation first (each client is "new" —a new situation is being moved into) without a diagnosis, respond to cues picked up along the way, and analyze afterward what happened. If there is a goal ("to help") it is like my goal at the circus: I didn't think about that once until I had arrived at Mr. Ringling. At that point, it had already been accomplished. In order to reach it, I had to "get in." The "getting in" was done step by step as things happened, with never a notion of what would or "should"—or even might—come next.

*Some Learnings from a Study of Psychotherapy with Schizophrenics**

Carl R. Rogers

Introduction

I began to feel, a number of years ago, that I had probably made the contributions of which I was capable to knowledge concerning psychotherapy with clinic clients. These achievements, modest and faulty though they might be, were gratifying. But I was eager to push out the boundaries of this hard-won knowledge. To what degree would our theories and hypotheses hold in working with more extreme groups? Specifically I was eager to test our theories in psychotherapeutic work with psychotic individuals, and with thoroughly normal individuals.

Two Basic Theories

Some years ago I formulated the view that it was not the special professional knowledge of the therapist, nor his intellectual conception of therapy (his "school of thought"), nor his techniques, which determined his effectiveness. I hypothesized that what was important was the extent to which he possessed certain personal attitudes in the relationship. I endeavored to define three of these which I regarded as basic—the realness, genuineness or congruence of the therapist; the degree of emphatic understanding of his client which he experienced and communicated; and the degree of unconditional positive regard or non-possessive liking which he felt toward his client.

The statement of this theoretical view almost immediately stimulated investigators to devise tools by which these at-

* Abridged from a paper in the *Pennsylvania Psychiatric Quarterly*, Summer 1962.

titudinal conditions could be measured. Significant studies have now been completed indicating that in an out-patient clinic population, greater degrees of these attitudes are indeed associated with more successful outcomes in therapy. This is true whether the measures of such attitudes are based upon the therapist's perception of their presence, or ratings by impartial observers, or the client's perception of the relationship. The client's perception of the extent to which these attitudes are present appears to be the most accurate predictor of eventual outcome.

A second line of work has been the investigation of the detailed process of therapy as it occurs in the client or patient. Based upon clinical observation a theory of the therapeutic process was formulated, and a Scale of Process was then developed which was intended to tap the changing manner and quality of the client's way of expressing himself as he moved in the continuum of therapy. In essence, the Scale endeavors to place any particular sample of the client's expression on a defined continuum. At one end the individual's psychological functioning is rigid, static, undifferentiated, impersonal. Constructs are fixed. He exhibits feelings but does not own them. He is remote from the experiencing going on within him. He is unable to relate. He is unable to communicate himself. He sees himself either as having no problems, or being in no way responsible for the problems which do exist. At the other end of the continuum the individual is functioning in a fluid, changing way, responsive to the everchanging experiencing which is going on within himself, and responsive to the events going on outside himself. Feelings are experienced with immediacy, are owned, may be expressed when appropriate. The individual is close to his experiencing and refers to it in guiding his behavor. Experience is construed tentatively, and new meanings are drawn from new experience. Relationships are real and fluid. The individual recognizes his responsibility for many of his own problems.

We set out to investigate the possibility that there is a lawful relationship between, on the one hand, the attitudinal conditions provided by the therapist, and the process of therapy and personality change in the client on the other. And we hypothesized that this relationship would hold whether we are speaking of therapy with schizophrenic, neurotic, or normal individuals.*

I want to tell you of some of the general and clinical things we have learned, and then a few of the objective findings which seem to be emerging.

The Low Probability of Change in Our Group of Hospitalized Schizophrenics

The first of these learnings is one which was gradually driven home to us. It is that we are probably dealing with one of the most difficult groups ever worked with, in terms of the probability of change. We are dealing with more or less chronically psychotic individuals. There is ample experience to indicate that this is a very difficult group. We are dealing with individuals whose socio-educational status is, on the average, quite low. We are dealing with individuals who, almost without exception, are not consciously desirous of psychotherapy, who tend to have the view that "talking can't help," and who do not see the therapist as a potentially helpful person. Some of them are not only unmotivated, but are actively antagonistic to any attempt to help them. When you consider these three factors together I believe my statement is correct that in terms of probability of change, this is one of the least promising groups ever taken on for psychotherapy.

In view of the low probability of change in our group, we cannot help but be impressed by another of our learnings, which is that we have very often been able to reach, to work with, and to facilitate some change in, these individuals. I will not expand that statement here, because I believe it will be evident in some of the other material I present.

The Special Problem of the Unmotivated Client

A major lesson we have gleaned is that the absence of conscious motivation constitutes a really profound problem in psychotherapy. This is certainly not news, but we have

* In the original paper there follows a description of the design and methods of the research in which eight therapists and forty-eight hospitalized schizophrenics—many of them regarded as chronic—were involved.

learned to recognize and formulate this issue much more clearly and sharply. There is a great difference between working with the consciously motivated client, whether neurotic or psychotic, and working with the person who has no such conscious motive, whether that person is normal, neurotic, or psychotic.

Working with the person who desires help is psychotherapy, as usually defined and understood. Not only is there a wealth of experience in this kind of relationship, but there has been a steady increase in empirical knowledge as well. I think that we know, objectively, a great deal about psychotherapy of this sort. We have a beginning objective knowledge of the kind of individual who is most likely to profit from it, the kind of relationship most likely to facilitate improvement, the kind of changes which most characteristically occur to a measurable degree.

But for working with the person who has no conscious desire for help we need, I believe, a new term. It is an area in which relatively little has been developed in the way of theories or concepts. Until very recently there have been no therapists willing to expose to view their way of working with these individuals. We owe a real debt to such therapists as Frieda Fromm-Reichman, John Rosen, and Carl Whitaker, for their willingness to demonstrate publicly and to record their way of dealing with these individuals, their way of provoking a relationship where no desire for a therapeutic relationship exists initially.

From our own experience in working with unmotivated schizophrenic individuals and a small matched group of unmotivated normal individuals I have come to a conclusion which you may regard as startling. It is my present conviction that working with a lack of conscious motivation in the individual is more difficult than working with the problem of psychosis. This is of course a subjective opinion, based in part on our general lack of success in trying to form a facilitative relationship with unmotivated "normals" of low socio-educational status. In so far as the two elements are separable, I believe the absence of conscious desire for help presents a greater challenge to the therapist than the presence of psychosis.

In any event, I have come to believe that we will make more progress in this area if we recognize that dealing with the person who does not wish help is a clearly different undertaking from psychotherapy, and if we build up the concepts, theories, and practices appropriate to it. We should not be misled by the fact that a relationship with such an individual may become psychotherapy when he chooses to seek help.

Some Observed Differences Between Schizophrenic and Clinic Clients

I should like now to turn to another area of observation, the great qualitative differences we have found between our schizophrenic clients, and the clinic clients with whom we had previously worked.

One of the deepest learnings in my previous clinical work had been the tremendous pull exerted in the client by the satisfaction of learning one's self. No matter how external the concern initially expressed by the client—the problem of his wife's behavior, or the choice of a vocational goal—once he had experienced the bitter-sweet satisfactions of self-exploration, this inevitably became the focus of therapy. I do not find this to be true with our schizophrenic clients. Even when we have established a relationship, even when the individual experiences some new facet of himself, and understands himself a bit more clearly, he does not necessarily continue along this line. For reasons I am not sure I understand, he does not find himself, except very occasionally, drawn to the exploring and experiencing of self. Instead, as Gendlin has pointed out, he is more likely to continue to externalize his problems, to refuse to own his feelings. Is this due to the nature of the schizophrenic reaction to life? Is it primarily characteristic of the chronically hospitalized person? Is it due to the low socio-educational status of our group? Is it simply that very few of our clients have reached the level of inner development where self-exploration is satisfying? I cannot be sure.

Having written that statement just a few days ago, I find it is now modified by objective facts. In a study in which the results were completed just a few hours before I left Madison, it was found—and this confirms my statement—that counseling center clients, on the average, showed a significantly greater depth of self-exploration than our schizophrenic clients. This was based on a new measure of intrapersonal exploration, developed out of our Process Scale and its derivatives. It was also found, in accordance with expectation, that the more successful cases showed an increase in degree of self-exploration over time, while the less successful cases—both neurotic and psychotic—showed actually *less* intrapersonal exploration later in therapy than they did early in therapy. But the surprising finding was that the more successful schizophrenic cases showed the greatest increase in depth of self-exploration from early to late, greater even than the successful neurotic cases. This is both pleasing and surprising. It means that in those schizophrenics who do show marked improvement on objective tests, this improvement is preceded by the spontaneous and feelingful expression of personally relevant material, by an active, struggling, fearful exploration of self. Put in very oversimplified terms, it appears in our sample that when a schizophrenic improves it is because he has entered into "therapy" as we have customarily understood it.

Another simple observation. Our schizophrenics tend to be either massively silent, or to engage in continuous (and not very revealing) conversation. It has been found that half of our schizophrenics, in their second interviews, show either less than 1% silence or more than 40% silence. This is sharply different from clinic clients. Our schizophrenic individuals tend to fend off a relationship either by an almost complete silence—often extending over many interviews—or by a flood of over-talk which is equally effective in preventing a real encounter.

Therapy as Relationship or Encounter

I want to turn now to learnings which are more personal, which have affected us more deeply as individual therapists. Perhaps the deepest of these learnings is a confirmation of, and an extension of, the concept that therapy has to do with the *relationship*, and has relatively little to do with techniques or with theory and ideology. In this respect I believe my views have become more, rather than less, extreme. I believe it is the *realness* of the therapist in the relationship which is the most important element. It is when the therapist is natural and spontaneous that he seems to be most effective. Probably this is a "trained humanness" as one of our therapists suggests, but in the moment it is the natural reaction of *this* person. Thus our sharply different therapists achieve good results in quite different ways. For one, an impatient, no-nonsense, let's put-the-cards-on-the-table approach is most effective, because in such an approach he is most openly being himself. For another it may be a much more gentle, and more obviously warm approach, because this is the way *this* therapist is. Our experience has deeply reinforced and extended my own view that the person who is able *openly* to be himself at that moment, as he is at the deepest levels he is able to be, is the effective therapist. Perhaps nothing else is of any importance.

The Place of Theory in Therapy

I feel that I have learned anew the appropriate place of theory—or more accurately, the appropriate absence of theory—in the actual conduct of therapy. As I listen to recordings of my own functioning with these schizophrenics, I can clearly note points at which my behavior is governed by my theory of therapy. As I have listened to a recording by Carl Whitaker of an interview with a psychotic, and discussed it with him, I note that some of his comments and questions—about cigars, for example—are governed by his theory of possible homosexual involvement in the patient. As I read the excerpts from Rosen's interviews in Scheflen's book, his theory of maternal rejection and incestuous desires as causes of schizophrenia often shows through. And as I listen to myself and these other therapists, I find I am not impressed by the relationship at the time these theories are evident. I suspect that each of us would be equally effective if we held quite different theories, providing we *believed* them. I would hazard the hypothesis that, in the immediate moment of the relationship, the particular theory of the therapist is irrelevant, and if it is in the therapist's consciousness at that moment, is probably *detrimental* to therapy. What I am saying is that it is the existential encounter which is important, and that in the immediate moment of the therapeutic relationship, consciousness of theory has no helpful place. Another way of stating this is that to the extent that we are thinking theoretically in the relationship, we become spectators, not players—and it is as players that we are effective. This is not to deny the fact that at some other time we may find it necessary and rewarding to develop a theory or theories regarding ourselves, our clients or patients, and the process of change itself. In the moment of relationship, however, I am inclined to believe that such theory is irrelevant or detrimental. So if theory is to be held at all, it seems to me that it should be held tentatively, lightly, flexibly, in a way which is freely open to change, and should be laid aside in the moment of encounter itself.

New Ways of Being in the Relationship

As we have been challenged by the difficulties of working with this group of schizophrenics we have learned new behaviors which are facilitative. How can we initiate a relationship with such individuals? One way which most of us have come to use is the expression by the therapist of his own immediate feelings, in a way which does not impose on the client. Thus with a very silent man, I have voiced such feelings as these: "I feel quite content with our silence this morning. Sometimes when we are not talking I feel sort of impatient and wanting to know what is going on but this morning it feels good just to be with you, quietly." Or with another client, "I feel that you are angry at me. I'm not sure, of course, because you haven't said so. But I keep wondering why you are angry with me." Such statements are expressed as inward musings, not as questions or demands. They may be much more extensive than these brief examples. They tend to put the therapist as a person into the relationship, without threatening the client by demands. They keep repeating in a variety of ways, "I am here. I am offering a relationship. I am a person with feelings, with perceptions. I am sensitive to you and anything you are willing to reveal about yourself." As Gendlin puts it, the therapist "can draw on his own momentary experiencing and find there an everpresent reservoir from which he can draw, and with which he can initiate, deepen, and carry on therapeutic interaction even with an unmotivated, silent, or externalizing person."

Some of our therapists go further in their behavior. One in particular is moving more and more toward allying himself with the hidden and unrevealed person in the schizophrenic, and openly "clobbering" the defensive shell. In his work there is a real similarity to Rosen or Whitaker. He is sensitively and obviously committed to the person who is in hiding, but he is quite violently and sometimes sarcastically critical of the psychotic symptoms, the fear of relating, the defenses and avoidances. Perhaps partly because this approach is congenial to him as a person, he is finding it effective. As we listen to the recorded interviews of the various therapists in our group, we are gradually broadening the repertoire of behaviors which are real for each of us in dealing with our psychotic clients, and are slowly hammering out ways of facilitating movement in the unmotivated person.

Waiting for the "Positive Will"

There is another lesson which we as therapists have learned, and that is the patience which is necessary to elicit what Rank has called the "positive will," in an individual in whom it has never been exercised. We have come to realize that almost none of the individuals with whom we have been working have ever been affirmed themselves. They have never, in any meaningful way, said "I feel", "I live", "I have a right to be." They have instead been passive receivers of life's hurts, blows, and events. It takes, in my experience, great patience to wait for the germination and budding of the will to say "I *am,* I deserve to *be.*" Yet the phenomenon of growth is in some respects all the more exciting because it has been so long dormant.

The Unimportance of Psychotic Content

Coming, as a number of our therapists did, from work with non-hospitalized individuals, the development of our feeling about the significance of psychotic content in our contacts is of interest. The simplest way of stating our present attitude is to say that we have learned how relatively unimportant is psychotic material. This could easily be misunderstood. The hallucination, the delusion, the bizarre language or posture has of course its significance in the psychological dynamics of the schizophrenic individual. But in the therapeutic relationship it simply forms a more difficult language of communication. Working with a child in play therapy, his language is his behavior, and this is often difficult to understand. Or in working with a foreign born client, garbled phrases and mispronunciations make understanding more difficult. So it is, it seems to us, with the schizophrenic. His scrambled incoherencies, his paranoid ideas, his hearing of voices, simply represent a mode of communicating himself which is often *very* difficult to understand. But we have found the difference one of degree rather than kind, and do not attach special importance, nor use special procedures when dealing with the person who is frankly psychotic in his way of expressing himself.

Some of the Objective Learnings

I would like to mention a few of the objective learnings which appear to be emerging from our analyses. I would, however, stress that most of these statements are very tentative, and may be modified by further study.

Regarding the Therapeutic Attitudinal "Conditions"

In endeavoring to measure the conditions of therapy as they exist in the therapist, we have found it possible to use three avenues of perception. Through using the Relationship Inventory, we have a measure of these attitudes as perceived by the therapist and by the client. Through using unbiased judges who listen to segments of interviews, we have been able to obtain ratings of the degree to which these conditions exist, as measured by an impartial observer. In this third mode of measurement it has been difficult to obtain suitable reliability between judges, indicating that it is not easy for anyone to be sure, from a therapist's words and tone of voice, whether he is genuine, whether he does really care. Reliability is somewhat higher in gauging the degree and accuracy of empathic understanding. The important thing, however, is that we have been able to develop means of measuring subtle and significant therapist attitudes, from each of three perceptual vantage points: from the viewpoint of the therapist himself, from the point of view of an observer, and from the viewpoint of the client.

Taking the client's perception of these attitudes we have an interesting finding. Those schizophrenics who perceive more of the attitudinal conditions of therapy in their therapist, as measured by their responses on the Relationship Inventory, show significantly more process movement as measured by the Process Scale, during the first 30 Interviews, than do those who perceive less of these therapeutic attitudes in the relationship.

Another preliminary finding has emerged, based upon the judges' ratings of one of these qualities, accurate empathy. This is contained in a study by Truax of eight schizophrenic patients—four of them showing improvement on psychological tests, and four showing some deterioration. Samples were taken from the first twenty-four interviews with these patients, 385 samples in all. Four judges listened to these samples independently, and rated the degree of accurate and sensitive empathy exhibited by the therapist. When the ratings were analyzed, it was found that the improving patients received consistently more sensitive empathy than did those who were not improving.

Our findings with schizophrenic clients also confirm an earlier finding with clinic clients. We have found that those schizophrenics who are less well adjusted, more psychotic, in terms of the MMPI, perceive less of these therapeutic conditions initially than do the better adjusted clients. Hence, it would appear that the perceived existence of these qualities in the relationship is an interactional thing. The client who is deeply disturbed has difficulty perceiving these attitudes even when they exist in the therapist.

The Measurement of Therapeutic Movement

The second important task of the research had to do with the measurement of movement or process in psychotherapy. It is clear that we have been able to develop a reasonably reliable, reasonably sensitive measure of process which can even discriminate differences between first and second halves of therapy interviews. This measure is based upon objectively discriminable verbal and voice clues. It has already turned up significant findings regarding out-patient clients, such as the unexpected finding that clients who are to be successful *begin* as well as end at a higher process level. This scale seems also to be applicable to interviews with schizophrenic clients as well as clinic clients, an achievement we had hardly dared to hope for.

The Relationship Between Conditions and Process

Our central hypothesis was, you will recall, that the more the therapeutic conditions existed in the therapist the more therapeutic movement, as measured by the Process Scales, would be found in the schizophrenic client. It is too early to say whether this hypothesis will be supported. Some of our early analyses tend to confirm the hypothesis, others do not. A confirming bit of evidence finds that the degree of movement in the first thirty interviews, as measured by the Process Scale, is positively related to the extent to which the therapeutic conditions are perceived by the client in the relationship. In other words, in this study, therapeutic movement is related to the client's perception of these qualities in the relationship. The more he perceives the therapist as empathic, for example, the more therapeutic movement he shows, and the more he changes. However, most of the correlations between judges' ratings of the therapist qualities and the degree of process movement in the client have not been significant. The most we can say at the moment is that a high degree of the therapist qualities appears to be associated with degree of positive change on psychological tests; and that therapeutic movement as measured by the Process Scale is also associated with positive change on psychological tests. The statement which would complete the logical triangle, that a high degree of therapist conditions is associated with marked therapeutic movement as measured by the Process Scale, cannot be made at this time because current findings are ambiguous.

Conclusion

This has been a personal report of a large venture, which is still far from complete. I have simply tried to sketch the outlines of our program with schizophrenics, and some of the meanings and changes it has involved for us as therapists. I have also tried to suggest the kind of findings which are emerging and will continue to emerge from our study.

There is obviously no conclusion possible, if I am speaking in objective terms. But subjectively it has had great significance to me that we have been able to reach these schizophrenic individuals as *persons*. We certainly cannot say what the final outcomes will be, and perhaps the degree of social recovery will be modest. But it has both personal and theoretical importance that we have been able to help withdrawn, bizarre, hopeless individuals become human beings.

I think of one man with whom I have spent many hours, including many hours of silence. There have been long stretches when I had no way of knowing whether the relationship had any meaning for him. He was uncommunicative, seemingly indifferent, withdrawn, inarticulate. I think of an hour when he felt completely worthless, hopeless, suicidal. He wanted to run away, wanted to do away with himself because, as he muttered in flat despair, "I just don't care." I responded, "I know you don't care about yourself, don't care at *all*, but I just want you to know that *I* care." And then, after a long pause, came a violent flood of deep, wracking, gasping sobs which continued for nearly half an hour. He had taken in the meaning of my feeling for him. It was dramatic evidence of what all of us have learned—that behind the curtains of silence, and hallucination, and strange talk, and hostility, and indifference, there is in each case a person, and that if we are skillful and fortunate we can *reach* that person, and can live, often for brief moments only, in a direct person-to-person relationship with him. To me that fact seems to say something about the nature of schizophrenia. It says something too about the nature of man and his craving for and fear of a deep human relationship. It seems to say that human beings are persons, whether we have labelled them as schizophrenic or whatever. And I know that these moments of real relationship with these real persons have been the essential reward for all of us participating in the program.

XI

> To separate the essential from the non-essential is what I call being spiritual.
>
> Franz Marc

In Hawaii during the war, women were told not to give servicemen a lift. I kept telling myself that I would obey this, and not take chances. But a difficulty was that a good deal of my driving was done in the country where there was no public transportation, there were few cars on the road because of gas rationing, and quite long distances between places. If I passed by a serviceman, he might have to stand there for a long time, waiting. Wasn't being in the military during a war bad enough? These men had been trained to kill, and soon they would be sent farther across the Pacific and would have to do it. Trouble enough. Suppose that were *my* son standing there?

No matter how often I decided *not* to stop, at the last moment my foot would go down on the brake, feeling as though it had been pushed by someone else. I didn't know then that the "someone else" was me, acting on my own authority and against the outside authority of warnings in the newspapers, from the military government, and from my husband—a preference for being human, at the risk of myself. If I did not yield to it, what value could my life have for me?

Usually I was not scared, but sometimes I was. Once, the soldier standing beside the road looked so very tough, his face pushed all out of shape as though he had been in fights, his mouth looking mean and ready for another. But still, he was human, and he was standing there alone. My foot went down on the brake as usual, and he got in the car. I thought perhaps if I talked fast enough he wouldn't have time to think of anything else, so I babbled. Suddenly he sat up in the seat and turned toward me, and I knew a moment of heightened fear. But then he said, "Wudja say dat again?" and I repeated what I had said, in which I had used some Hawaiian words without noticing. He settled back easily into his seat with a sigh and said, "I jus' *love* de sound of dem Hawaiian woids."

I don't think he was an angel. I think he probably had done all the things that I thought he had. But still, in that moment he *was* an angel: his deep human tenderness and love of beauty came through so completely, so *real,* that he made the poets whom I knew seem suddenly like phoney copies of themselves, as far from *real* as the man who saw in a Rorschach blot where most people see two men, "The shadows of the silhouettes of the ghosts of two puppets."

". . . there is in each case a person."

I have never known it to be otherwise. (In poets, too!) My schizophrenic friend is sometimes violent—kicks out doors and windows, frames and all. He is also cunning—he does this in places where he knows that he will not be picked up by the police. He does not fear jail, but he fears that he might be sent to a mental hospital and robbed of a part of his life that is very dear to him. A psychiatrist recently recommended that he commit himself to an institution "but I faked him out by telling him I would go home and to my psychiatrist there." His cunning has sometimes the look of ugliness—when he comes to me for money "to go see Larry" which is not what he will use it for. His eyes look like squirmy worms then. But what is this but an expression of how he feels about himself when he asks for money and tells a lie to get it? Shall I live with his fawning, cheating—or with how he feels about it, the meaning of it to him? Other times he *is* a frolicking angel—as humorous and saintly as he looks.

Which "he" is *he?*

One evening with him I always think of as perfect because there is never anything in it that I wish to change. Everyone else had gone to bed. He started painting at the dining table. I sat beside him. Somehow I was moved to take a sheet of his paper and begin painting too. The paintings were my own —and unlike his. They were also unlike anything I had ever done before. The *way* that he went about his painting communicated itself to me without words, and I have ever since thought of him as my "teacher" although he made no attempt to teach. At the same time, he got something from me which he has ever since thought of as therapy.

"And I know that these moments of real relationship with these real persons have been the essential reward for all of us. . . ."

What madness is it that a psychiatrist can write that

> "We psychiatrists measure an individual's behavior against his socially-determined role or roles (as we see them), and on this basis decide which behavior is incorrect or aberrant and *if so how much.*"

To me, this is not *psycho* (soul) therapy but manipulation into phoniness. One friend of mine who was psychoanalyzed five times between the ages of 17 and 53 told me, "Each psychiatrist takes you all apart and lays out all the pieces on his desk. Then he puts you back together the way that *he* thinks you look best." Another who was a patient in an army psychiatric hospital for two years said, "Psychiatry is a shoehorn to get you into a shoe that doesn't fit."

That is the old way (prevalent, though with exceptions). Psychotherapy seems to me to be one of the most *moving* (on the move, and in the direction of sensitivity and humanness) fields of work that we have today—in contrast, say, to education, philosophy, medicine, and academic psychology. I have heard of many of the "old guard" who have modified their view in recent years, and moved in the direction of "adjustment to myself" psychology (or autonomy) and of none working in self psychology who have left it for the "adjustment to society" way. I don't know how many are working at this time in one way, how many in the other. It is not necessary for me to know this. The *direction* in psychotherapy today seems to me to be clear. A hangover from the past, represented in whatever present numbers, is still not contemporary—does not reflect *what is going on now.* The process is the same whether it is within a profession or within myself: whatever rigidities I still have from the past are unfortunate and I regret them, but my *living* is in the direction of breaking them up and being more me—less of what I have been taught or told or have heard so much that it has become a part of me, and more of my own direct observation, my own noticing, past and present, and moving in my *own* direction.

When I am aware of this in myself, I can recognize it in others too and unite with their movement—what is going on now—rather than conflict or collide with what they haven't got rid of yet. Or more precisely, not letting what I haven't got rid of yet collide with what they haven't got rid of yet. Or as one psychologist told me when I chided him mildly for making statements which were his *training* and were not in accord with what he had discovered for himself, "I've broken out of the shell, but bits of it are still clinging to my feet." I had thought of myself as having got (largely) out of the mud but it was still sticking to my heels no matter how much I tried to shake it off.

In the process of emergence, I become horrified by the cruelty which is a kind of killing, that is done through ignorance. It shocks me that I have been a part of this cruelty and that to a degree I still am. At first, this was so terrible to me that it seemed unbearable—like pain going on and on until it reaches the point of explosion. I feared this exploding in myself—complete disruption—chaos. How could I live with my knowing? It seemed to me that perhaps it was better to be ignorant, and I tried desperately to shut out what I knew (was seeing for myself). "The truth shall make ye free, but first it shall make ye miserable." But in spite of my desperation, my strongest efforts in that direction seemed to me feeble: I had gone too far to turn back. My own tide was sweeping me on.

What could I *do?* What could one small I do against massive ignorance which I had only recently begun to get free of myself? I felt the way that I did when I went to Boston and thought, "I am never going to be able to change Boston."

But there is always something that I can do.

I can bring my mind into the immediate present, and to the realm in which I *can* be active, and let myself act. That is, let myself be responsive. I accept my own limitations, and do not try to do what I can't. To try to do what I can't (in the literal sense) fills me with frustration, even if I am only trying to figure out how I can do it. This last, in fact, can so paralyze me that I can't do even what I *can.*

When the Navajos were in great distress all over their nation (it is that to them), what I could do seemed so little, so utterly futile, that I got caught in "What's the use of doing *anything?*" A drop of water in a sieve. This was not exaggeration. In point of fact, when the government got around to distributing a couple of million dollars among the Navajos it came to about $30 a person—to people who needed shoes, blankets and food, and were already in debt to the trader. Their food was pitiful. A bowl of mush, and picking the tiny bits of meat still clinging to lamb bones after the rest had been eaten was a meal. In the cold harsh winter they slept in their hogans with only a sheepskin between them and the ground. They had not only the winter to contend with, but also *us*. We had reduced their sheep to prevent overgrazing and then blizzards decimated those which we had permitted them to keep. We had imposed on them a senseless world which had no meaning to them. When a Navajo caught some horses on Wild Horse Mesa, the government branding inspector refused to brand them because there was no bill of sale. And still, they never lost their sensing of the humor of it. One Navajo said, with recognition completely free of malice, "Oh, if we could run the white man's world for three days, what a mess we would make of it!" Successive Commissioners of Indian Affairs, each with his own values, had imposed these values on the Navajos. During one commissioner's term, education was most important. During another's, health was most important. Under one commissioner, it was most important to send the children off the reservation to school. Under another, it was most important to keep the children in school on the reservation. The Navajos were "sold" on these and so many other changing values that when one of their own young men came back from theology school and wished to help them, they told him, "We are sorry. We like you, and we would like to do as you say, but there's no more room in the head."

What could *I* do? I could be human. *Real* human.

How would *I* feel if people from another culture were in power and insisted that I must adopt their ways without question, that everything I believed and lived for was wrong, and then one person from the race listened to me, understood my world and let it be?

I could stand beside a Navajo who didn't know English,

while I knew no Navajo, and share his feeling-world with him. When I did this, I had the feeling of increased strength in me, and it seemed to me that there was this increase in him, too.

I could sit with a Navajo woman whose infant had been run over by the wheel of a cart. She sat on the earth floor of the hogan with the infant on her lap. Its thigh was so bloated that it seemed to belong to a pig rather than an infant. The wound gaped so that you could look right into the distorted flesh. The child suffered silently. The mother suffered silently. I could join this silent suffering and share it, not giving advice because what advice was there to give? Somehow this silent sharing eases pain. It seems unlikely that an infant could take part in this and yet, when we are disturbed doesn't this affect the infant? So why should it not affect the infant too that mother and visitor knew this quiet in themselves?

"And I know that these moments of real relationship with these real persons have been the essential reward. . . ."

When I was in the hospital dying very rapidly—I could feel so much less life in me each day—all my successes and my failures seemed to cancel out. It was the same with the good that I had done and the bad that I had done. They canceled out. What was left—all that was left—was the wordless resonance, a feeling of love that was universe-al, a kind of singing intelligence that was universe-al too, that knew all and was undismayed. That was all that was left—and that was all that I had ever wanted. I was sad, with water in my eyes, for all the not-living that I had done.

XII

> If the matter is one that can be settled by observation, make the observation yourself. Aristotle could have avoided the mistake of thinking that women have fewer teeth than men, by the simple device of asking Mrs. Aristotle to keep her mouth open while he counted. He did not do so because he thought he knew. Thinking you know when in fact you don't is a fatal mistake to which we are all prone.
>
> Bertrand Russell*

In my beginning, mind and body functioned together easily and I knew no difference between the one and the other. I did not think of them as "mine"—to be used by me.

When I was two, there were things that I knew I must not do: I would be punished (don't like) if I did them. But I was not punished for doing what I had not been *told* not to do because "How can a child know that he shouldn't do it unless he has been told?" These were facts known to me in my two-year-old fashion, although the way that I express them now was impossible for me then.

One afternoon my mother slept and I explored. It was a kind of casting around or scanning until something attracted me. I moved by this being drawn to, and things combined themselves with me. There was a large ball of twine on a chair in the dining room. From my close-to-the-floor position, the dining table was posts with a roof on top. I put the string around a pillar and went on to another—back and forth, around, across, myself crawling around and through with the ball of string, feeling both the weaver and the woven. I was

* From *Unpopular Essays*, by Bertrand Russell. Simon and Schuster, New York, 1966, p. 103.

happy with the happenings of which I was a part, with no separative thought of myself as the doer. My concentration was total: I had no thoughts *about.* All my senses were alive (alert, functioning, experiencing) in the present moment, receiving sight and smell and touch and sound and (putting the string in my mouth, or my tongue against a table leg) taste.

At other times I "thought about" but I was not yet trapped or confused by language and knew these thoughts as going on in me, not something that "I" did. All that I received through all my senses moved in me as in a kaleidoscope, falling into a pattern and then, when that pattern was complete, falling apart and then into a new arrangement—dropping out some pieces and bringing in others that had been left out—and I was the interested observer of thoughts going on in me. One of the good times for this was when I had been put to bed and was alone, away from everyone—although the sounds of others in the family, muted by doors and distance, non-interfering, non-demanding, were reassuringly present. I rested, floated, drifted, with awareness, until I was comfortably swallowed by sleep.

For two decades afterward, all this still happened in me easily. No matter how much I might have got lost in other worlds, I still could find my way back to my own (the one I was most happy in, that felt like me) and to the same kind of concentration or reflection. When I was twenty-one, I read a book while knitting and while another book was being read to me, receiving both, and still receiving all the sights and smells and sounds of what else was around me—the foggy night air, the buoys moaning from the bay, the music of the voice reading the words to me, the colors and shapes within the lighted room—and very much aware of *me* in the sense of what was going on in me as the alive and vital receiver and integrator of all this—although what I had come to know as "I" was the integrator only because all this was happening in me. At the same time, reflection was going on, making patterns, ever-changing, out of all that was received, each one seeming to deepen the understanding of the others.

But the person who read to me objected to the book that I was reading with my eyes. He said that I couldn't be listening to the one that he was reading. So I stopped that. And the "I" who stopped that began to seem more "me."

For so many years it seemed to me a waste of time and life to do only one thing at a time. When I worked for a society woman when I was fifteen, I licked stamps and put them on envelopes while she gave me instructions. She said that I wasn't paying attention to what she said. Like school.

In my early thirties I had so far lost touch with this that when my ten-year-old daughter bounced a balloon while I was talking to her, I asked her (accusingly) if she had heard what I said. She seemed so intent on bouncing the balloon. She said yes, and—while she still went on bouncing the balloon—repeated everything that I had said. I was recalled to my own experience and felt ashamed that I had been suspicious. I could at least have been matter-of-fact, genuinely inquiring.

Whether a person is listening while he is bouncing a balloon depends on *how* he does it, which can only be known inside himself because that is where the *how* is happening. When I was knitting, reading and being read to, at the end I corrected the reader's mispronunciations matter-of-factly, without criticism. He was cross and said that I couldn't be listening, receiving the meaning of what he read, if I were thinking about mispronunciations. But I wasn't thinking about them. They just went on, like the ticking of a clock. He was right in saying that my doing this bothered him (a fact of himself), wrong when he insisted that he knew what was going on *in me*. When things like this were said to me often enough, I became confused about myself.

To my present self it seems extraordinary, even unbelievable, that I could have done all these things at once when I was young, even though I know that it is true. But I have read several times recently in books by psychologists and psychiatrists about elderly women in England, of no particular education and with little money (I like those words better than "low socio-economic status") who listen to the radio, knit, and chat all at once, without losing track of any of them. Who knows what else may be going on in them at the same time?

It is very difficult for me now to arrive at the stillness (in me) in which all things happen. I cannot do it when others around me are not still. By still, I do not mean "making no noise" or "sitting still" but having the stillness inside so that nothing is forced, exaggerated, or full of intention—the stillness of birdsongs or coyotes' voices which I hear with no sense of interruption. They make no demands of me. Sometimes I can be with other people in this way. Then I make no demands of myself. I just am. I get this same sensing from the mountains which are still and from the oceans which are never still. Each in its own fashion *is*. This links with something in myself, brings into awareness a depth of myself with which I have been out of touch. It seems to me that when we indiscriminately wipe out the rest of nature, as we so much have done and progressively are doing, we lose touch with something in ourselves. When I am aware of myself as a part of nature—rocks, trees, birds, earth, air, coyotes, lice—I am most human, and most released to human activity. Myself is freed of me.

They are what they are: I am what I am.

Life is ease-y then: I feel at ease, at rest in a lively active/passive way. Writing about it recalls me to it: my stiff old body at this moment is free. I have let go of me. My body moves with a feeling of fluidity. There is no pain. I have the feeling of joy that I knew when I was young. These are the reasons why I would like to be young again. Otherwise, I prefer where I am, reaching the other side of confusion and coming into the blue again. But I do not consider the confusion *necessary* except in the context of our place and time —our ignorance.

When I was small my explorations went on both outside me and within me, united in the same questioning. Later my questioning turned more to outside, less to what was going on in me. Because the questioning went on within my skull, I didn't notice that in most of it *I* was left out: like the argument between my father and my uncle Bob about money, referred to earlier, which went on in my head but had nothing to do with *me*.

Partly I got twisted around in this way: when I included myself, sophisticated people rebuked me. In a discussion of philosophy with a professor I said, "Do you mean that if I did this, then that would happen?" I said this to make clear my understanding of what he had said, to see if I had got it right, and at the same time to check it within myself to see if it was in accord with me. I was told with annoyance and condescension "You always make it personal!" "You can't argue abstractly."

This happened to me very often.

How could I accept or reject the abstraction if I had not checked it with myself first?

But I felt ignorant, stupid, female. And also confused. It seemed to me that the professors were talking somewhere out in front of their faces, like the balloons in cartoon strips. Their talk seemed unreal, rootless, disconnected from ourselves and consequently having nothing to do with anyone else either.

Van Dusen writes of the above, "This is very meaningful to me. Abstract talk seems to me not only 'in front of the face' but not even in this room or this world. It is like rumors about something real."

But the only way that the professors would accept me, to the extent of permitting me to talk with them, was to do it their way. A great deal of what I read was written in that way too, establishing a habit in me, in spite of the fact that it seemed to me that an awful lot of it was nonsense. I have seen so many graduate students who felt this way about what they had to do, but they made themselves do it to get their permit to work, and came out of it believing—and taking seriously—what in the first place they knew was nonsense. Gossip and rumors. That's what it is.

I think this sort of thing happens to most of us in one way or another. We get all twisted and distorted and then we say that's what a human being *is*. It seems to me that we don't know and never can know what a human being *is*—in the sense of "can be." Just as with a child we can only watch him grow, unfolding like a flower, with no knowing ever what he may become. "Here is this infant in my arms. I love her. I am full of curiosity about her." As she grows, I am enchanted to observe her growing, in some ways like all other children, in others so very much herself. When my first child was punished even mildly for doing something she had been told not to do, she did not do it again. When my second child did not follow the same pattern, I did not understand him. I looked at him and thought, "He's intelligent, and he's got honest eyes, so he should come out all right"—and waited for the time when he had a command of words, so that he could tell me about himself. When he was four years old, I told him about this way in which I did not understand him, and he said (not having got the hang of punctuation yet), "I'd rather do what I want to do and get spanked than not do it."

Watching, waiting, while at the same time enjoying both what is here now (this child) and where she has been along the way, with the future always full of questions. When I do this with her, I also do it with me, and with all the human world—the human race. It is like being a child again with the whole future before me, through which I find my *own* way by my choosing. Today I choose this, but tomorrow? How can I know? Something I have not yet seen or known may present itself. Or a change may have taken place. For thirty years, nothing would have been more incredible to me than that ten years later I should be doing a book with a psychotherapist. But then there were changes in me, and in some psychotherapists and psychotherapy.

What other changes may be taking place that I haven't latched onto yet, that I might like to move into when this book is done? The thought comes to me sometimes, and some possibilities have shown themselves to me. I don't know yet what I shall happen into next. It isn't yet time. There is something that I haven't finished yet, which I wish to complete before I move on. I felt that way when I was dying very rapidly twelve years ago. It wasn't yet time. I can explore dying more fully later on. While I live, it is life that I explore. I feel fairly sure that this is also the best preparation for dying when it comes. One moves then easily from one into the other, not clinging to what one should (for one's own happiness and the relief of others) give up. Throughout life it is the same, that I must always be ready to receive and ready to relinquish, in accord with the time.

I am arriving at an understanding of my knowing that I must live with life, that this is the problem. It seems to me futile to study the problems of infants, the problems of children, the problems of the first child, of the last child, of the middle child; the problems of youth, of middle age, of the menopausal age, of geriatrics; of the poor, of the rich, of the middle class, of minority groups, and all the other problems. If all that time, energy, intelligence, and money were put into a study of the problem of being human, all the others would be resolved. There's just one problem—a very big one, but still just one.

When I lose touch with life, and the life and goings-on in

me, I can trust myself to regain it through my dreams, which are good advisors.

When I was small my dreams, even when I was terribly afraid of them, were helpful. When I acted on what they told me, I didn't have them any more.

When I first went to school, I was exposed to tickling by other children. I fell to the ground and writhed in agony. At night I was afraid to go to sleep because of what I dreamed. In my dream, I was lying on bare ground in a clearing (there were no sidewalks in the village where I lived) and out of the woods that surrounded me (there were woods in the village) would stream in single file half a dozen ugly little dwarfs (the way the children looked to me when they tickled me). They were heading straight for my body. I knew with horror that if I moved the least little bit I would be discovered and they would pounce on me, torture me, and I wouldn't be able to shake them off. So I lay *utterly* quiet—not a quiver, made myself dead. Then they changed their course to swing around me, beyond my head, and streamed off into the woods on the other side. When the last one had disappeared in the woods, I was safe.

I do not recall that I made any conscious connection between this frequent dream which was always exactly the same, and my being tickled, but I learned that absolute inner stillness was something I could arrive at in my waking life when someone began to tickle me. Then I felt nothing when I was "tickled," and the children said, "Heck, you're no fun!" and left me.

This stillness was useful to me in other emergencies, for many years. When shock threatened to disorganize me, I could arrive at immediate inner stillness and do whatever needed to be done. Later, I lost touch with that, chiefly through my strong and forceful husband who insisted that I must *think first*. When I had got into the habit of doing that, then the only way that I could still chaos was to make myself still, hold myself together, use force on myself to keep my *exterior* calm, instead of arriving at the stillness which *is* in me, as I think it must be in all of us. Within this stillness *I do not think:* thoughts occur to me if thoughts are needed; otherwise, I *act,* without thinking. This comes out much better than when I think first, or "try to figure things out."

When I was sick and couldn't get well, and all my money

had got used up and all my efforts to correct both conditions failed, and I lost faith in me, I arrived at this stillness again in another way—while awake—and it has never entirely left me since. It was a different image this time, one that I have since seen described by someone else. But the image itself is not important, or whether it agrees with that of anyone else. The importance (to me) is in what it says: inner wisdom showing me something that in my panic I had overlooked.

Trust in my dreams is something quite new to me. My early conditioning said that dreams were nonsense, to be laughed at and ignored. My later conditioning said to mistrust them (and myself) and then I did what a young woman did recently in telling me about her recurrent dream: "There is a mound of dung, with a bright shining jewel on top. The feeling is good. The dung doesn't smell of anything. I like it. I feel good with it. . . . I suppose what it is telling me is that I'm just a pile of shit." With that *good* feeling?

In actuality, this young woman had been *told* by her conventional family and friends that she was a horrible person because of things she did. She was made to feel dirty and foul ("smelly"). Her own opinion of herself shows in her dream—what "should" be smelly and awful was pleasant to her.

My experience of dreams since I have learned to notice and accept them is that they are accurate and honest, utterly without deception, speaking the only language that they know. To understand them I must not twist what they are saying but accept it, sink into it, and let them speak to me.

Frank Barron writes of a dream he had:

> The dream was one of almost total darkness. Gradually in the darkness I discerned shadowy figures, almost lost so shadowy they were: first one or two, then a few more, then by the tens and by the hundreds, all dimly bodied forth. Soon I realized that I could see them only from the waist up, and that from the waist down they were caught in a mire, an oozy swamp which was gradually sucking them downwards and out of sight. Then I saw that each of the figures was holding on for dear life, left hand and right hand, to the hands of two other persons in the mire. Suddenly the dream became a nightmare, for in an instant of recognition and realization I

knew that each one in the swamp thought that with his right hand he clasped the hand of someone on firm ground, and with his left hand he clasped the hand of someone deeper in the swamp, but that actually everyone was stuck to the same extent in the swamp and each had the same illusion. Then I recognized the dim figures in the dream, one after another; those lost figures were psychologists and psychiatrists all, and in the light of day they were connected with one another in some such wise as in my dream.*

The last sentence gives Frank Barron's own knowing of the dream, his recognition of its meaning to him. Then he writes, "The state of affairs in psychotherapy (even in such a hotbed of self-examination with the help of someone else as Berkeley, California, was in those days) was never quite that bad." It seems to me that the dream was more accurate before a part of it was rejected. Whatever I add or take away from what is given—added or taken away by what I think of as "myself"—seems to reduce accuracy. Many psychologists and psychiatrists lean heavily on one authority or another, not standing on their own, which to me is a mire in itself. This may come about in part through the academic notion of "giving credit," which is a bondage. If I am forever giving credit to someone else, I tie myself to the other person's thinking, and am unaware that all creation (including mine) *is* free. At the same time, by giving credit to someone else I avoid responsibility. Any criticism or discredit that may arise can be referred to the other. This is illusion, for in fact what I do with someone else's thinking becomes my own. It stands or falls as *mine,* and is *my* responsibility. What anyone does with anything that I have written is *theirs,* not mine, and is their responsibility. They cannot shove it off on me.

As for the vast unknowns about people that seem to be indicated in Frank Barron's dream, I think this is true. I sometimes want to throw out this manuscript because of all that I do not know, that is not known to the "experts" either. I remember a time when I urged Carl Rogers to make his knowings more available to people in general, and he replied

* From *Creativity and Psychological Health*, by Frank Barron. Copyright 1963, D. Van Nostrand Company, Inc., Princeton, New Jersey, p. 69.

that he did not feel that anything he knew was sufficiently proven for that. I told him that this waiting for proof is what hangs us up all the time; that we're never going to arrive at anything *final*, and so it seemed to me best to make available to everyone what is or seems to be known, for them to use in working things out for themselves, each in his own fashion, not as a "follower" believing in outside authority.

Van Dusen says of Frank Barron's dream, in a letter, "It comments very well on our sure scientific knowledge. The more I examine it, the more doubtful it becomes. If investigator A believes a thing, he gets positive experimental results. B, a skeptic, can't get the same results. Conclusion: what we find in the world rests on our beliefs. Like Wittgenstein. The world as it is, is unknowable unless one takes as world one's own belief system."

Werner Heisenberg writes:

> We have to remember that what we observe is not nature in itself, but nature exposed to our method of questioning.*

That is a deep level statement, for some not easily grasped. At another level, not saying the same thing but leading to it, I heard recently of some research in which the design of the experiment seemed to be without flaw to the scientists who studied it. It was an ESP test of the Rhine type, with cards. The people tested were separated into the "sheep" and the "goats." Those who believed in ESP scored better than chance. Those who disbelieved in ESP scored lower than chance.

Why should I *believe* anything?

Why should I believe *anything?*

Is it necessary to believe, or is this just an old concept that has been getting us into difficulties throughout history and in our present personal lives?

It seems likely that I shall soon go to New Mexico to explore for a place that would be good for people to grow on. I am attracted to this, and now feels like a good time to do it. It feels "right" (good) although intellect can make no sense

* From *Physics and Philosophy*, by Werner Heisenberg. Harper & Bros., New York, 1958, p. 58.

of it. A few days ago I thought, "I don't really believe that I'm going to find it." That made it seem senseless for me to go. Then I cast around inside myself and came upon another thought, "I don't really believe that I'm *not* going to find it, either." For a moment, the two together seemed "wrong" —to leave me in a sort of nothing or nowhere, and that of course is supposed to be a very bad place to be. But then I found that I liked it, realized that it is a very good and comfortable (and accurate) place: I neither believe nor disbelieve. I am attracted to New Mexico at this time, and I am going to find out if the place I want is there.

This "nowhere" or "nothing" (no belief) is not at all dismal. In a mild but happy sort of way, it excites me. Everything is open. I'm curious. Along with this, there is openness to anything else that comes along which might lead in the same direction, to what I want. What I want is known to me in a clear although largely unspecified way: I cannot communicate it to someone else, just as I cannot communicate to someone else any other creative urge until it has happened—a book, a play, a painting. *Before* it has happened, I have to invent something with my intellect (in order to get a grant or other support) which cannot possibly be what will happen of itself because that is always beyond my knowing or my imagination. It is the same with a business when it is gone about in the creative way, for the creative is not *what* is done but *how* it is done—just as it is in my living. To "live with life" I must live in the creative way. All other "ways of life" are really the same way, although they may seem very different. That is true of the creative way, too. This is what makes it so easy for me to stumble, to become confused. The same words refer to such different modes of experiencing, according to the *how* in me. It is the *how* in me that I must notice.

My inner direction at this time is very clear to me. It is when this inner direction is most clear to me that I am least able to explain my actions to anyone else—unless *they* are close to *their* inner direction; then we understand each other and nothing needs to be explained.

First I must get together with *me*. Then, although my outer course may seem erratic and incomprehensible to others, it is in fact a steady one in me. Without this centering *first,* the "same" kind of wandering (as seen by others) is not the same. The impetus does not come from the same place.

When I have the openness which accompanies the inner direction, the bumps or obstacles which may appear along the way are not something to be shoved aside or overridden —or permitted to block me—but are something to be guided by. When my approach is that of a river finding its course, I arrive at my sea, or destination. The idea "New Mexico" is not so much a plan as it is a place to begin, or a way to begin on beginning. I might not even get there, if something else appears along the way. I am not lost; my inner direction, which can be known only to me, is clear and strong, the way it was when I was young and, leaving my family, took off on my own. I knew no one else who had done what I did then, but that was not *why* I did it. It arose out of *my* life, my *own* experience, and was most agreeable to others too, although that was not *why* I did it. It was not agreeable to my parents because they thought that I "should" do something else but in fact, in time, it worked out well for them too. I have never regretted it, always have known it as a rich, full and free and very responsible time in my life.

Later, I became muddled, and wondered why living, which had been easy, had become so difficult and wearing and I had the feeling of being stuck or stopped or bound—or grinding on.

It seems to me that this way of living is also the way of exploration and the way of science, when it is truly exploring and truly scientific, and the way of learning (education), and that this is also the way of infants before they know anything except what they are born with. "An adult is a deteriorated child." (Max Wertheimer)

If the paper which follows is read without belief, either the belief which is believing or the belief which is disbelief, then I think that something may come of it within each reader, although for no two persons will it be quite the same. When my interest is in what each person gets out of this or any other journey for himself, I am not dismayed that they didn't get out of it what I did, do not try to force them to see *my* view, but am enchanted by how different the same scene looks when seen through different windows. At the same time, it clearly *is* the same scene, and I—and they—do not feel isolated and alone.

The Natural Depth in Man

Wilson Van Dusen

> One of the major theoretical hypotheses of client-centered therapy is that during therapy the concept of the self is revised to assimilate basic experiences which had previously been denied to awareness as threatening.
>
> Rogers and Dymond[1]

Introduction

For a long while it has seemed to me that there is a natural depth in man's experience that is little known to the outer social man. I will wander in this depth to show something of its unique ways, language, and outlook on the outer world. Client-centered therapy is one means of letting come forth some of the personal meanings from this depth.

Let me set the context of this wandering. It seems to me that Carl Rogers entered the house of man through the door of science but went straight toward the idiographic, the unique experiencing and describing of a person's world. I move in the same direction, but I enter by the back door of phenomenology. It is my hope that those who observe something of the strange contours in man will appreciate the appropriateness of the client-centered method which encourages the uniqueness of a person to come forth as a person in the light of day—often for the first time.

When I say that Carl Rogers entered by the door of science, I mean he has brought with him respect for what can be measured and publicly validated by the communal methods of science. The Rogers and Dymond[1] volume on the testing of the client-centered method is an example. But even here he chose one of the most idiographic of statistical approaches, the Q sort. He and his colleagues produced numerous healthy correlations as contrasted with the sicker data (to scientists) of personal meanings in the varied landscape of personal worlds.

In my personal encounters with Carl Rogers and in his filmed psychotherapy sessions, it is apparent that here is a man who easily and naturally enters the world of other persons and enjoys their company, their diverse personal worlds. The client is suddenly not isolated in his uniqueness but finds an understanding companion in Carl Rogers. It seems that many of the so-called non-directive or client-centered therapists have copied only the outer shell of Rogers' work. As an example, two pseudo-trained non-directive therapists came to dine with my wife and me. They practiced their subtle art of reflecting everything back to her. It annoyed her because it was an artificial game by which they remained free of even meeting us. It is remarkable that the attempt to emulate a man who easily walks in the world of others should have such an opposite result. In religious terms, this is known as acting by the law rather than the spirit.

There is a growing confluence of people in psychology and kindred disciplines who move in the direction of humanism while respectful of science. Rogers is an example and so is the loosely integrated and somewhat diverse group of existentialists and the real Adlerians. I expect and hope the confluence will be called humanism. The confluence can be as large as the humanities themselves. Science and humanism have different aims, though they can be productive brethren. Humanism can supply root questions and root speculations of man together with raw data of human experience—phenomenological descriptions. The science of psychology can be considerably enriched by this data; yet the aim of science is toward the communal, common, consensually validated—what is general or lawful and can be measured and shown consistently by publicly agreed upon methods. On the negative side, science can see humanism as vague in method, inconsistent, and unsure in results. Humanism, on the other hand, can well see in science a great monster that eats up the test protocols of 1,000 people to spew out one or two generalities. Science digests uniqueness and turns it to the generally valid. Humanism can gain from science methods which would increase the validity and reliability of observation, while science can get from humanism the varied, provocative and question-awakening richness of individual experience. These should be two sides of the same coin.

There are variants between these two. The case history approach, especially when it translates into a Standard History Form, produces partly-digested human worlds. ("This is a male white protestant, 32 years of age, with a high school education.") It lies between the unique and the communalized. The schools of psychology are sectarian variants. They translate the unique world of an individual into an agreed upon frame of reference. For instance the classical Freudians would discover sex and aggression in any person and explain present behavior by reference to psychosexual development. Like churches, these sects communicate well to their own brethren but look askance upon those of other creeds. Schoolism in psychology is quite common, and to anyone who has dealt with church groups it looks to be a remarkably similar phenomenon. It seems to me the schools are saying largely the same things in different languages from different viewpoints—sort of a modern Babel. There are taboos between the scientific and humanistic camps. It is a serious violation for the humanist to pretend his observations or opinions of man have any relationship to science. This can lead to modern civilized forms of the cat-o-nine-tails to be laid on. It is less of a taboo violation for the scientist to pretend he is a humanist. Invasions from this direction are common.

Prior to the work of scientific psychology, there is a need to find the human experience the science is to deal with. This corresponds to geologists picking up stones and paleontologists finding bones. Usually human experience is bypassed, or squeezed into a data processable form by getting the responses of people to a protocol (i.e. "answers to these questions are a measure of introversion"). The protocol may involve responses to a fixed array of lights or other equipment. The person is translated into a fixed pre-determined protocol which is then data-processable to arrive at some general statements of people. The protocol is fashioned to note certain given aspects, although many protocols don't even do this reliably or validly.

The individual's life, spread through time, and different under different circumstances, is the original complete protocol. In this form it can be barely read, let alone statistically processed. Moreover, it is a protocol in its own unique language having a unique frame of reference. It is important to recognize at the outset that the phenomenology of an individual life is not scientific and only with some work can be processed into the mode of science. A human's life I see as pre-scientific, although it may provide data for science as stones do to a geologist. I once set out to describe the world of a schizophrenic man.[2] Much of my description consisted of quotes from the man himself, for in some lucid moments he captured better than anyone what was happening to him. I deliberately avoided the categories of the usual case history. What was important to him went into it. For instance, the meaning of the things in his pockets was important to him. To check on the validity of this description of his world I could but ask the subject of this schizophrenic world—himself. He found it accurate except that his world fell in pieces and my description was coherent. He lacked the continuity of meaning we take for granted. There were deadly places of lostness in which he wasn't sure who he was, where he was, or what it all meant. To more accurately describe his world I would have had to use a poetic form of language with horrible blank spaces—something in the mode of Joyce's *Ulysses*. Validity in science is what can be commonly agreed upon. Validity in humanism is finally the subject himself. Science finds the common in the varied. Phenomenology finds the varied in the common.

Carl Rogers combines both spirits but by nature he is closest to humanism. In his work with individuals he is closest to a phenomenologist who can enter and participate in the unique world of another person. His "laws" are statements of major facets of the lives of people. Some scientists are critical because these statements are hard to measure, data process, and prove in this way. But if the venture of humanism does its work well, it may enlarge and enrich the data science works upon. Ideally, it would enrich the very methods of science, which after all, change with the times.

A Venture to the Interior

I set for myself the task of describing the inner nature of man. In fact, I describe myself but with the expectation it has some generality to the worlds of others. An adequate description of the subject matter requires some entering into its very mood rather than playing at being an objective, detached scientist. If I remained detached I would be akin to a fundamentalist minister trying to capture the depth of an African tribal celebration in fundamentalist terms. We will see that the very language and the frame of reference of the innermost is quite different from our ordinary way of thinking. I would have no quarrel with anyone who asserted the language of the novelist, poet, or musician is closer to the quality of human experience than the language of psychologists.

A rough map of the territory may help clarify. One can distinguish what is outermost in man as that which is more or less perceived by and available to himself, from an innermost that extends from this to aspects of himself which are quite unlike the public self. Each of these areas can be roughly ordered. The real outermost is the world which we more or less agree upon and assume exists regardless of ourselves. One can experience trees, sky, people and all "others" as though they are a world superordinate to us. It exists. Less outermost are perceptions which are relative to the perceiver. For instance I see this room, my study, as relative to me. An empty study divorced from my presence and use of it would be something else, strange to contemplate. At about this level might be included the reaction of one person to another in which there is awareness of both as present, i.e., "Your words surprise me." At the inner border of the outermost I would include sensations of one's own body. It can be like a thing possessed and used. For instance I notice a little stiffness here and there and need to change position. It is almost as though I tolerate my body as a necessary other. In this outer region I would also include states of mind in which I think of what to say to others, so to speak —adjustments of my mask.

The inner me (or to borrow the French l'autre moi, the other me) begins at this border and extends to puzzling and even unknown aspects of myself. Near the border of the outer, l'autre moi includes spontaneous associations of thought which arise unbidden when in a social context or alone. Inside that are vaguely felt sensations, feelings, images in something of an apparently hodge-podge order. There are states in which one can awaken the inner life to stand forth of its own. There are, in rough order toward the innermost, drifting-fantasy, auto-hypnotic inner perception, the hypnogogic state (the territory near sleep or relaxation) and dreaming while asleep. There are also so-called abnormal states such as hallucinations, altered perceptions of the body and outer world, visions, etc.

The whole of this inner me, this inner territory, is marked by one distinct difference from the outer me. The life of the inner arises spontaneously, showing surprising twists and turns. The outer me is guided by circumstances or what I choose to do and think. The border of the inner is reached when spontaneous thoughts, feelings, or images arise, perhaps related to the outer situation at hand, but still autonomously surprising in their nature. The surprise may be a little one, where one can see the associative link. Or it may be like a dream where the message of the inner is not really understandable. To the conscious outer self the inner, or l'autre moi, is marked by this spontaneous unaccountability. There are lovely images in literature where the relationship of the inner and outer are portrayed. For instance, in the Bible there are references to the relationship of the servant to the master of the house. It may come as a surprise to ego, but in this imagery ego is servant and the inner is the master. The master provides and the servant makes do with what is provided, i.e. the inner springs provide resources and the outer is judged by what is made of them. Some creative people make good use of the inner and are judged favorably by themselves and the world.

Upon entrance into the inner world, it is appropriate that the language one uses broadens into imagery and a more freely literate style. To the scientific outer ego it might be appropriate to say, "I perceive a flattened affect." But to say, "Everything seems flat and meaningless to me as though the world is an abandoned cemetery" is closer to an accurate representation of the inner. As will be seen, the inner thinks in terms of images. In the innermost, feelings, images, thoughts are all one living varying process. Here is an example which illustrates the central theme of the inner.

I see (in the sense of conceive, which is a weak perception!) an image such as an area of light surrounding an eye-shaped area of darkness. This is inner rather than fabricated because it arises of itself. Its meaning or use is not now apparent. If I forget the image, its use can be permanently lost. Insofar as I invest in the image, I can gradually *feel* out its *meaning*. It comes slowly. Something about darkness seeing, but I don't understand it. Ego can react and reject the idea as illogical. Darkness cannot see—it is a foolish bit of imagination. But in the inner world conceptual images grow slowly, like a seed, from almost nothing to more and more differentiated forms. Let us see.

In the above image the eye is black nothing outlined by light. It is seen as something in contrast with the light. It is seen. The associations run to a variety of scenes such as a person appearing dumb in the light of others' wisdom. But it is more than that. Ignorance has a primal sense of itself out of contrast with what is not. Suddenly the black eye has light in it. This reflects ignorance being aware of itself. I have a funny feeling of ignorance's wisdom, and the eye winks at me. Then out of my ignorance I see that I am ignorance aware of itself surrounded by light, paradoxically being light insofar as aware of my darkness. The image is not foolish imagination now. I can see it is an accurate representation of my state at that moment. I can see that I do not understand.

In a similar way I can wander through my own feelings. The language is again that of images. From the perspective of this mood the scientific way of presenting human matters looks like the naked timbers of a house constrained to rise vertically, evenly spaced, rising just so far under command. My mood wanders like water over ground, down through dark soil, creeping around rocks deep into unseen places, wandering endlessly to the sea as if possessed. If the inner spirit wanders thus, why imitate it in any stiff, pale, reflected image?

My feelings come to a pause like water slowly collecting and spreading in a low place in the ground. It grows stiflingly hot and feverish, but what is my concern? Hot blind concern not yet able to name itself—fearing it may end in useless emptiness like water that can't bring anything to fruition. In the dark fingers of self lies nameless concern, its own creation of pressure. The feeling changes as though right has been done by naming it, permitting fidelity to itself.

There open out fragments of images. Let's respect these too and see where they lead. I see an incense burner in a church (tears at naming it), old hand-wrought metal, a fragrant woody odor that fills with its presence. There are fragments of Hebrew on parchment. The sense of lovingly carved wood and stolid images looking back at me. The sense of stillness, endlessness; the sense of the sacred and holy. Now the pen stops, for in this peace there are no queries and nothing to be done. It is its own doing.

Standing back and evaluating as ego I'd have to say that I entered my feelings, first finding them random and wandering, then gradually centering in concern, and finally coming to a peaceful state which is familiar and welcome.

Are not the layers of self always in the same sequence? First stands forth straight-timbered ego consciousness which obeys the will of worldly custom smiling to please the customers. Behind this the wayward wandering spirit which is at home in nameless feelings and intimations. Having entered this dark and uncertain passageway, in a moment there come forth images, memories, groups of possibilities. If the ego consciousness smiling at customers, and obeying custom is me, then somewhere beyond feelings rises l'autre moi—the other me in a strange land of exotic markets and places where foreign tongues are intuited and nothing is strange. Like an intermediary, feeling lies between me and the other me. It is the intimation of the deeper self. I can hardly say my feelings are not me, yet they are so little of my direct fashioning and follow so little my customs that I cannot claim to be their master but rather their servant. They are the go-between, the mediator between me and the other me.

The idea of l'autre moi, the other me, accords full rights and even personhood to the inner. The other me is what others have called the unconscious. It must have been ego consciousness which named the other region (un-conscious) as un-like itself. Out of this prejudice has come the professional cult of *un*doing the *un*conscious that might rise to respectable egohood. Might it not be that consciousness can become less the storekeeper and more comfortable in its own home? Might it not be that the inner is the real master and the better master?

For some years I've examined what occurs in the hypnogogic state—the state between waking and sleeping. Here is one's personal psychological laboratory that can be experimented in endlessly without funds or approval. It is the experiences in this state which have done more than anything else to reinforce my conviction that the inner world differs in critical ways from our ordinary consciousness. I would suggest that anyone interested explore the same region. People apparently vary in their power to even note the hypnogogic experiences. The region borders sleep; many slip into a dream and are asleep dreaming before they notice any transition. It takes some concentration to hover near sleep without falling asleep. Under careful examination it soon becomes apparent that a delicate balance must be struck between ego consciousness and the autonomous appearance of l'autre moi. It is easy to try to produce imagery or try to think of thus and so and hence block the appearance of l'autre moi. One can wander for months in a region where worries or concerns fill the mind and ego consciousness frets over its problems. Or one can still the mind by fixating on one thought or one sensation until it is relatively quiet. Then around the fringe of the conscious center spontaneous images or words appear. Whether words or images come first reflects personality differences (intuitive versus sensation types in Jung's terms?). These fragile fringe phenomena are an unexpected, delicate penumbra surrounding the conscious center. Note that ego consciousness needs to be stilled, and out of the way, for the hypnogogic to appear. Efforts of consciousness to make images appear, easily blocks them. Most of the hypnogogic experience is not readily remembered. It takes extra work to put it in storage. Occasionally an image will break through clearly and be rememberable. Often the hypnogogic experiences are just out of reach of what can be remembered unless they are immediately recorded, which action tends to disrupt the state. Sometimes, when awakening from sleep, the hypnogogic experiences are relatively strong and dream-like, fading as one enters wakefulness. One can prolong this in-between state. It is possible to allow the hypnogogic experiences to grow strong and then deal with them without blocking them out. It is then possible to address and deal with whatever images or voices appear. Mine consist mainly of things said in a voice like my own thoughts. For many people images are more common than voices.[3] What is said or imaged always rises unexpectedly and usually is not even immediately understood. This is one difference from ego consciousness. When I think of something I have some anticipation for the next thought, and it appears familiar and understood. The hypnogogic is unanticipated, strange, surprising, and not even in one's frame of reference. My earliest experiences were often only a single word or brief phrase said in an instant of time. My reaction was to grab onto the words, becoming too conscious and delaying the reappearance of the hypnogogic. For instance, I heard "stairs" or "my liberal art class." To my knowledge, Silberer[4] was the first to note that these thoughts are autosymbolic. *Autosymbolic* refers to the mental process by which the mind easily forms representations of its own state. If one examines what was in consciousness at the moment the hypnogogic image arose, its meaning is clearer. For instance, I was centering my attention on what it meant to rise in thoughts when the word "stairs" appeared. I was thinking of the richness of imagery when I heard "my liberal art class." At least on one level the hypnogogic appears to be spontaneous images of one's innermost thoughts or state at that moment. It isn't the same thing as dreams for my dreams are entirely in scenes with almost no words.

Silberer describes three types of hypnogogic experiences. (1) Material, a thought is represented by a picture (or as above with words). (2) Functional effort phenomenon, i.e., the effort to stay awake is represented by a picture. (3) Somatic, i.e., a deep breath is seen as a table lifting. Some examples will clarify. When I should be getting up I've sometimes heard the alarm clock though it had not rung. Once I heard, "Don't think I go on any more," as I was awakening. I took this to mean the inner process would be cut off. Once, when I slept more than needed I heard, "Drunk, drunk," which captured the groggy feeling. I have when I've slept too much. Once, while meditating near sleep, I heard, "I sent you the howling jacket. I sent it by mail." The phrase had no meaning to me until it was carefully studied. At that time I was coming under a great deal of public scorn which felt like wearing a howling jacket. The scorn arose over the mere possession of a plant that had been sent to me by mail. But then, who is the I that claims to have done this? Once when listening to random (and unremembered) voices in this state I heard, "Drowned people in the brush." The "drowned" fitted with my barely apprehending what was said, and they were invisible or in the brush. Once I centered my attention on a headache and heard, "Non-material, there is an old bitch or bastard and you can't see how intimate he is. Concentrate on the pain. It is useful." While meditating on an answer to a question, I picked up "Don't get lost over pennies," which seemed to say don't get lost in minor aspects.

Sometimes one can address the process without disrupting it too much. Once I asked who speaks, and heard "Advertisement." When I asked again, I heard "Edward Connes," a name which means nothing to me. This is an example of another process occurring. Sometimes I pick up strange words like "anzeema" or a name like Jenkins. At the moment the word formed I could see that it is an exact representation of my feelings at that moment. One can watch words and sounds arise around feelings as though they take the form of feelings and image them in a most intimate way. Perhaps the hypnogogic or similar state might permit a fundamental study of the matrix of language itself.

These experiences have several characteristics. They arise without anticipation. They are usually not understood immediately, just as though one is suddenly shown an image or told something by another person. Most, if not all, of it is symbolic. Most, if not all, appears to be a representation of the innermost thought or feelings. It occurs in a kind of timeless world which is now without past or future. It borders memory and usually requires some extra work to fix in memory. Most importantly its frame of reference is the innermost-subjective-now reality of the individual. *It is as though, at our core is a process which cannot help but accurately represent itself.* I am in accord with Carl Jung when he says that dreams (and now the hypnogogic) are from the objective unconscious—objective in the sense of truthful and accurate regardless of any misconceptions we may have of ourselves. Yet l'autre moi, though it can have many identities like a shifting myth, has no one personal identity such as I have.

Some will wonder at the similarities of this region of normal experience and the hallucinations of schizophrenics. There are both consistent similarities and differences. Hallucinations reported by patients are much louder or more visible than the majority of hypnogogic experiences. A phrase said in the normal hypnogogic is as faint as a whispered notion, whereas most schizophrenic auditory hallucinations are as loud as a normal spoken voice. Things seen are often indistinguishable from reality. Hallucinations often have no fixed identity or a slippery changing identity similar to the hypnogogic. The normal experiences are usually momentary while hallucinations can keep up for years. Both states appear to be different from dreams, i.e., a person tormented by a set of hallucinatory tormentors will usually have no sign of them in dreams. Hallucinations are unanticipated, often symbolic, often not understood and seem to have the same frame of reference, i.e., the innermost subjective of the person. I have chosen patients who can distinguish between themselves and their hallucinations. I strike up a relationship both with the patient and the hallucinated figures. Then it is possible to question the hallucinations directly, and the patient reveals their response. In this way it is possible to do a psychological examination of voices and the patient as separate entities. Curiously, hallucinations seem to be sicker than the patient on psychological tests! One patient's hallucinations said they were from Washington, D.C. and showed him a U. S. sign. Through the patient I asked were they from the city of Washington, D. C. near Virginia or were they at the seat of the government of the patient. They affirmed it was the latter. I have a good deal of evidence, too lengthy to cite, that hallucinations have this same inner subjective frame of reference. The patient is often fooled by it because it sounds as though the hallucinations refer to the outer real world when they are speaking of the person's own subjective world. This man's hallucinations felt they were his government. They commented for years how he was in their power. Hence they are from his Washington, D. C.

My guess is that psychotic hallucinations and hypnogogic experiences are related but not identical. Hallucinations are stronger, more durable, and more consistently organized. They are often more alien and negative too. It may be that these are one process made different by the patient's poorer insight and greater struggle against l'autre moi. For instance one woman had loud voices shouting at her all day so that she killed one person, broke windows, etc. Nothing abated the voices so I helped her to meet and deal with them. As she came to accept what they suggested, they quietened, became less negative, and eventually merged into her thoughts.

It would appear that the intensity of most schizophrenic hallucinations, as contrasted with the faint delicacy of hypnogogic experiences, is a function of the patient's struggle against l'autre moi. It looks as though hallucinations are generated by the patient's rigid alienation from his other side, making it more attacking, fiercer, and alien in its aspects. There are less common positive, supportive hallucinations which are also alien and archetypal in nature. One man had a spirited female hallucination who was gay, humorous, and quite wise in matters of ancient symbols in marked contrast to the patient's disinterest and lack of ability in the matter of symbols. In fact this lady hallucination was the most gifted and sensitive person I've met! Hallucinations seem either lower than the patient (meaner, more sexual, more limited and tormenting) or higher, but either way alien. They may be either higher or lower because they are the rejected, unused aspects of the person.

A dream is a normal and common message from l'autre moi having less interference from consciousness in its creation than any of the phenomena (except hallucinations) touched upon thus far. An example will remind the reader of its strange language and strange frame of reference. Last night I dreamt I saw a little something sticking out of my breast over my heart. To my horror it moved, making me feel that something alien was alive in me. I pulled it out, fearing it may mean there are more in me. It looked like an ancient creature one inch long, vaguely like a cobra in shape. It had a tail webbed like a duck's foot. It moved and seemed very ancient.

At first the dream meant almost nothing. Then it came to me. "Out of my breast, from the region of my heart," I associated with writing this chapter. I pull something out of my heart. It is more living than I wanted it to be. Like the cobra it is lovely (my association to *my* cobra—yours is a different creature), rhythmic and yet deadly in its possibilities. The deadly aspect of this paper won't be apparent yet, but it has to do with ego's limited scope, being servant to an inner master. What I pull out of me is ancient, before man. It is conceivable that this inner world was latent before man and probably the first world known by man. It wiggles and frightens—this ancient knowledge I pull out of my breast—fearsome for there is more there that I have not yet met and understood. The dream images my feelings today. It was written in an ancient language I am only coming to understand. Emanuel Swedenborg, in the mid-18th century, aptly called this the language of representations. It was given from the region of mind that represents itself, endlessly. Swedenborg appears to have presented the most detailed study of this language ever attempted.[5] As shown in his *Spiritual Diary*,[6] most of his work was done in a state similar to the hypnogogic. It is his contention, backed by much evidence, that this inner language of representations has as its reference innermost subjective qualities, and is the language of much of the Bible. His material is so extensive that scholars have taken a lifetime to fully comprehend it. Seen in this light, much that seems obscure and archaic in the Bible takes on great meaning.

Implications of l'autre moi

There are relatively consistent differences between what can be identified as ego consciousness and l'autre moi. Ego anticipates its acts and appears to choose them. They have an immediate meaning to ego. In contrast l'autre moi seems unanticipated, surprising, and the choosing is not apparent. To ego it is more or less alien. What ego does has immediate meaning to itself. The acts of l'autre moi appear meaningless or difficult to translate into usable meaning. The majority of people in the world discard messages of l'autre moi as useless, unimportant (how many study their dreams?), or even non-existent. Meaning to ego has reference to things of the world or reference to me. Meaning to l'autre moi exists in itself. Ego acts in a temporal landscape with past and future. Within the view of l'autre moi there is timelessness. Everything is now. Now is the whole scope of things. Ego has a single identity. I am Joe Doakes. L'autre moi can be said to either have no identity or all identities. The two seem to be the same. Though invisible in itself, it can represent itself in all things. Ego sees itself as an agent in an independent real world. L'autre moi is itself its own agent-world. From the viewpoint of l'autre moi all worlds are representations of itself. Ego must search for what is true and carefully check on it. L'autre moi represents truth instantly and without effort. After years of work with people's dreams I have far more respect for the unfailing accuracy of dreams than I have for what people say of themselves. All considered, it seems l'autre moi is wiser than ego. There are two aspects of the other me that are difficult to see from ego's limited view. L'autre moi sees the world as representative of itself. In dreams, friends may be used as examples of one's own traits. The same is true of houses, cats, etc. From the inside view, all things are examples of aspects of this larger self. I've received many humorous comments from l'autre moi when awakening, where the world was taken as its representation. For instance I once heard someone say, "here is a mondo (problem in the Zen sense) for you." In the next instant I was awake looking at the problem, the world itself. The real world is the unsolved problem presented to me. L'autre moi also uses time itself as its representation. It is as though it uses time as one might use the sequence of a drama to illustrate the character of a person. Its aim seems to be to portray its nature with a time sequence as a device to do this. Its time-transcending quality seems to permit the occasional appearance of precognition. It is as though the past, present, and future were all one to it.

All of these make fairly substantial differences between a person on the surface and a person in depth. One could devise a fairly adequate boundary between these regions. The ego seems to choose its thoughts and acts. L'autre moi is given. Ego understands its meaning, but l'autre moi speaks a symbolic and somewhat alien language. The boundary is crossed when ego fashions a fantasy only to find that the fantasy develops surprising changes in a life of its own, pointing beyond the little scope of ego. The schizophrenic who has approached this boundary from the other side often finds it difficult to locate. For instance, some who have recovered from manifest hallucinations that faded into their thoughts are then in some doubt (and are sometimes technically delusional) over which thoughts were theirs and which were given by other forces. The more I look at the hypnogogic process in which thoughts dawn out of themselves the less I am sure there is a valid boundary. The thoughts that come into my head also seem to come out of nowhere. I cannot grasp or really know how my thoughts are thought. It is clear that thoughts arise out of feelings, and one has little to do with fashioning feelings which are the supportive background of thought and key to its meaning and use. Feeling seems to be the root out of which thought comes.

When examined phenomenologically, it isn't difficult to undermine the primacy of this independent agent called ego, self, personal identity, or whatever. In the ordinary world of events it isn't surprising that words can issue forth endlessly as evidence of an agent self that makes itself known in this way. Yet, in the inward scene, the use of thought is not a simple matter. I hear these words coming one by one, and hear less certainly the next words and have the barest presentiment of words even a dozen or so further advanced than what is in consciousness now. From the phrase in consciousness one can go to the fairly well intimated next few words to the barely implied. I am lost in a sea of further presentiments of thoughts, feelings, images further beyond. If I take as my identity this floating in midstream of consciousness at this point, I can call unknown and even awesome the above stream of past and even more so the below stream of not yet. But why stream at all? How do I make these thoughts?

In the hypnogogic state it is apparent that thought can arise spontaneously of itself. The round of thought becomes even more mysterious when one tries to stop thinking. Then it becomes apparent that the thoughts we credit to our fashioning can ride roughshod over us and go about their business. Attempting to stop thought awakens a sea of images and potentialities within the person. At best one can stop this consensually validated, directed social intercourse and thereby lay bare the underpinnings of thought. One can look under the mask of socially-sanctioned communication at the source of thought and see its more primal and autonomous nature. There are individual differences even at this level. For some, feeling becomes image, for others it becomes words, for others the inner is like an imageless, wordless drama known before word or image. It is a delicate process, the mind watching itself at work. If the whole of the inner process differentiates into thought and this thought is ego, then ego would be the fixed conception of a changing process. Ego would be a result of l'autre moi. Add to this the accumulation of memory and the tie to a slowly changing body, and one has ego. The more one becomes acquainted with this primal underpinning of thought, the less it appears really different from agent self—except that it is more gifted and richer in conception. Ego is like its poorer brother, strayed from home. The real difference is that the inner is pre-eminently self-representational while normal thought is much more estranged into relating to, describing, dealing with everything but itself. I don't know if the inner process always knows itself to be self-representational, nor do we often recognize that we criticize in neighbors what are our own faults. They hurt us in neighbors because they are close to home in ourselves. So again, the difference between these areas is little. Rather than reduce the inner workings to the little ego, I would prefer to relate little ego to the wealthier, freer, more creative, deadly-accurate inner workings.

Ego and l'autre moi can be seen as one or split into two. Often ego overlooks the wild utterances of its other side. Worse yet, it may fight against them. All psychopathology seems of this nature. What is not accepted as truly one's own returns to plague one as a psychic or psychosomatic symptom. It becomes a thorn in the side, a thorn made thorny by one's rejection of it. Schizophrenia is an example of the revolution of l'autre moi. It musters force and descends on hapless ego, buffeting it with hallucinations and misperceptions so that vaunted ego can hardly find the real world anymore. The chronic schizophrenic is a sad and comic picture of a tiny ego claiming victory while all the while pronouncing the very words of l'autre moi. One schizophrenic aptly saw a giant beast wearing pants like a man, having in its long snout-jaw a tiny man which it endlessly tossed around, neither spitting man out into the world nor swallowing him in death. Such a monster I can appreciate and respect. This was an accurate picture of this man's state—a hapless victim of a monster. This is the end result of the long road of the little ego that cannot meet or learn from its other side. The servant who thinks he is master is made more servile. The servant who shares in the wealth of his master enjoys peace. The ego self is a small segment of self, merely the self of public places in a world that is mostly private.

If ego and l'autre moi are seen as one, then the conception of self is enlarged. In spite of extensive training to express one's self in a language, one would need to accept an inner native ability and predilection for representing one's state in another primal and even archaic language. In this older language endless images of the world or phrases of the world language are used to faithfully represent one's state.

Here the difficulty lies. Can I speak a language not taught me, a rich language, faithfully showing myself even when I am relatively lost and ignorant? And how can I do this effortlessly? To do this one has to let go of a conventional picture of the self to permit this effortless, wiser one to stand forth. Were I a primitive, I might pray to the inner one, seeking its guidance in signs. But, as a psychologist, I stand in awesome respect for its richness and its spontaneous wisdom. It is appropriate perhaps simply to notice, describe and respect what lies beyond the bounds of struggling ego. Gabriel Marcel[7] describes a mystery as what transcends its own data. Certainly this other side of us transcends the data given us. Is the innermost faculty a symbolic mirror which can but represent? This faculty deserves respect since it lies at the threshold of what we know ourselves to be, and it implies more than can be understood.

There is much more that could be said, some of it requiring many images and dreams to substantiate. One is that this inner self seems to have different values from the ego. What these inner values are cannot be described easily though they seem to involve inner truth, inner worth, and inner uses that transcend ordinary values in the world. Certainly the inner can be more critical of ourselves and our motives than we are inclined to be. I recall my joy at completing a masterpiece of a Ph.D. dissertation only to find my dream describing it as a fair painting but done in drab browns and greys (which I confess to be true).

It is best to leave open the question of the real nature of l'autre moi so that the matter will not be set and sealed by the relative ignorance of ego. But for those who would wish to venture further, I would draw attention to the extensive writings of Emanuel Swedenborg on this subject. Writing in the mid 1700's Swedenborg examined the language of representation (images) both as it arose in himself and in the Bible.

> Representations are nothing but images of spiritual things in natural ones, and when the former are rightly represented in the latter, then the two correspond. Yet the

> man who knows not what the Spiritual is, but only the Natural, is capable of thinking that such representations and derivative correspondences are impossible, for he might say to himself, "How can what is spiritual act upon what is material?" But if he will reflect upon the things taking place in himself every moment, he may be able to gain some idea of these matters; for instance, how the will can act upon the muscles of the body and effect real emotions; also how thought can act upon the organs of speech, moving the lungs, trachea, tongue and lips and thus produce speech; and also how the affections can act on the face and there present images of themselves so that another often thereby knows what is being thought and felt. These examples give some idea of what representations . . . are. (Arcana Coelestia, paragraph 4044).

For Swedenborg what we term the inner, subjective, l'autre moi is the spiritual in a man with a natural faculty to truly represent itself. It transcends and has corresponding to it, the ego or proprium, in his terms; just as the spiritual of man transcends but has corresponding to it, the gestures of the face.

The World as Representative

I would like to return from the inner subjective into the world of people and things, but this time prepared with some insights from l'autre moi. My diplomatic mission is to end the struggles of ego, the island of personal identity, that it might open out into the world which is its representation. The death of ego is the birth of everything else.

In my work in a mental hospital fate sends to my office a very mad fellow, a big handsome Greek with graying hair. He loudly proclaims that something must be done to get rid of the "niggers" (who are death itself). They are controlling minds, a crazy thing for there is no mind, he says. They "sex" young boys (who don't exist in the hospital) and by this monstrous practice control others and pollute the atmosphere. He wants my help to get authorities here to wipe them out. Over the years he has devised many ways of partly controlling their tortures. There must be a wad of papers over the left pocket, blinders on his eyes to shield him from the radiation of others' eyes, a wet rag on his left shoulder, wet hair under a hat, certain buttons undone, shoes unlaced, etc. Now and then, when he is under grave stress of these bastards, he asks me to go to town for a certain brand of uncontaminated vitamins, all of which he swallows in two days. Somehow the vitamins throw the Negro gangsters off the track. My friend Joe is always angry and disgusted at the pollution he sees around him. Clouds are blankets of dangerous radiation (what fools that the world can't see this!). Every leaf is sickening with cancerous growth. His own head is rotted, and they do ghastly injuries to his body (which I am sure he feels). For all his unique views, Joe impresses people as an intelligent, socially-perceptive person who is capable of sly digs at hospital personnel. He once was delighted to beat me at a test of intelligence. In spite of 17 years of hospitalization, he could go out to the free world if he could just shut up about this crime. But Joe goes on fighting for his rights and goes on being denied them because he insists. There are some 880,000 Joes locked up in the limited existence of mental hospitals. One thing is clear in my years with these people—the more mentally ill they are, the more they are caught in egocentricity, selfishness, and uselessness. In mental hospitals there are many relics of people with vaunted selfhood even though their world is riddled with strange influences. The more vaunted, the more riddled. It is apparent to us, but not to Joe, that this diseased world is representative of him since the disease walks where he walks. He contends that it occurs only in the hospital, but when he visited a large city, there was this same monstrous business going on. It is the same thing whether we say the man is diseased or his world is diseased. The two are really one.

But this example is far from the world of the reader because you perhaps aren't so mad. In ordinary social intercourse with others or in the more intense experience of group psychotherapy there is a manifest difference between the person who must explain and must defend his views from the person who enjoys and easily accepts the worlds of other persons. The first is a bore, a difficult person, one who rouses anger or needs to be treated carefully. The second is an easier, more likeable person. In group psychotherapy one often sees the clever professional or layman who must tell, must be right, must interpret what the other person really means. A less obnoxious egotism is the one who is very concerned what to say, how a thing will be taken, whether he is understood or appreciated. These people are painfully trapped in the evaluations of others because they are so concerned for themselves. In my own little world it has become clear to me that I am of use to others insofar as I am not concerned for my reputation. I've seen brilliant intuitive insights come out of me when I was only aware of the other person. I've also seen miserable failure when I was out to prove that I was clever. I look for pride in *me*, and if it is there, I'd best be silent. In both therapy and social relations there is an almost mystic beauty when my self vanishes and the style, nuances of bodily movement, voice inflection and concerns of the other person fill the whole stage. I am liked best when *I* am not.

Ego or self awareness seems like a pitiful little cramp, like a bit of food stuck in the throat. In a Zen novel a wise monk, confronted by an assassin's sword, admired the rainbow colors gleaming on its edge as it came. The assassin didn't kill him because it was so obvious he was no ordinary man. The beast, personal pride, is most difficult in workshops when one is a leader before many people. Once I wanted to beat my old teacher, Fritz Perls, at dream work before a group. Luckily I was miserably poor and he was most masterful for all to see.

Egotism is plain unhandy. Insofar as I am Somebody, I cut myself off from other-buddies. I am trapped in should/oughts, moment by moment weighed and evaluated in the eyes of others, a victim of opinion (mine and theirs), constrained to put on my best shopkeeper's face and coat. One becomes caught within imposed goals. "Everyone should try to be successful, have a big car, color TV, etc." Ego or pride is separateness, and separateness conflicts with how to be *with* others. Insofar as one is separate, one has a castle to defend, and there is defensiveness. There is a Chinese saying that "Locked doors make thieves." The one that is not permitted in the little cell self escapes into the world (is projected). "Those monsters who would start war overseas!"—but I've been known to start a war over the breakfast table. Sometimes the good side of a person is projected too. For instance, the one who marvels at the cleverness of others often projects greater greatness than exists in the others. Adolescents (young and old) are famous for this. Life is a kind of long-term seasoning where it is discovered that there isn't greater wisdom or stupidity out there than there is in here.

There are many ways that this so-called boundary between self and others is crossed. Love is one way, friendship another. In love it is as though the self and the loved one are a one which cannot easily be separated. Any relaxed (ego-less) perception is an example of this crossing. Watch, listen, let the other come forth. Enjoy the lovely nuances, the moods, feelings and concerns of the other one. Death to self is the birth of all else. When an emphatic therapist is with a patient, there is little sense of self. The mild self present is mostly the other one. The weighing, measuring, worrying of one's self is undone by opening into the life of the other person.

This can be true even with material things. I can be wrapped in the "howling jacket" of my concerns, or let the things before me speak of their nature. Outside my window is an oak tree. Every time I see it, it says something different. Today a woodpecker works on its side which is riddled with woodpecker holes. Peck, peck, peck—it says impatient. Not thorough. Other days I see its strong limbs cast eternally in a dance posture saying "thus" like a Hindu symbol of endless meaning. Now the branches hang down leafless and say "go to earth, peaceful." Such an eloquent tree. Its massive trunk goes to earth and says, "I am real before you." And so it is. In this relaxed perception of things each wordlessly reveals something and endlessly changes in nuances of meaning. Obviously the eloquence of the tree is mine. It says my meanings. Others see another tree there. Dull wits see merely an oak tree. In the relaxed, congruent, empathetic, unconditional regard for things they become beautifully eloquent of their/your nature. The perceived and apperceived world is one's own representation. Any meaning in the world has one's own signature on it. Insofar as this deepest nature of the world is found, one's own deepest nature stands forth. A Zen monk was pouring flax seed when a young monk asked what was the Buddha nature. The wiser one answered, "Three pounds of flax."

This is not merely a figure of speech, an allusion. If the flax were examined for months, leisurely, emphatically, the Buddha nature would gradually come forth and be seen. Peck, peck, peck. We are too impatient for this.

In effect I am saying the little enclosed self or ego can open inwardly into its own wiser aspects. Or it can open outwardly until it is nothing but all things. One can be friend to and enjoy all of one's selves! It is difficult for ego to give up its little defended citadel. Ego doubts the wisdom of its inner aspects, such as dreams or fantasies. The less it knows of it, the more it doubts it. Ego is afraid to open its door outwardly to others. How might they hurt me if they knew my real feelings? And, why should I pay so much attention to them?—they are so dull and routine! Others are dull insofar as they are unknown. It continually amazes me how interesting anyone can be insofar as I open to understand them. And even these interesting ones become dull whn I quit paying them heed. They become routine and dull again. The same is true for the material world. The major difference between one who enjoys art and a clod is that the first opens himself to appreciate what is before him. Did I ever tell you of the beauty of cobwebs on a dirty window? But that is another long story.

Little ego is concerned for its individuality. Insofar as it is little more than the reflected opinions of parents, friends, and society, it is little more than an average, a cheap common item on the market. The more interesting individuality, such as that of a tree, lies in every aspect in which it is unique, one of a kind, different. It may seem paradoxical, but the more an individual permits expression of his uniqueness (becomes individuated in Jung's sense) the more universal social values (Adler's social interest) emerge. It is the repeated experience of psychotherapists to see these core social values emerge with the fuller expression of a person's potential. The ego can overlook the sufferings of people rejected for their skin color. The inner cannot, for these people, too, are its representation. Insofar as the individual allows himself to be as he is, he is inclined to allow others to be also. This is simply another aspect of the way all others are one's own representation. We can accurately measure what goes on in the deep recesses of the inner world of a person by measuring what he does with others. It is more than a curious fancy that dreams represent one's relationship to aspects of one's self by relations to others in dreams. One is not simply the image of the other. They are one.

Individuality, in the sense of a defended citadel in the midst of a sea of others, is hardly worth having—an encumbrance best overcome. Individuality is of worth insofar as it means allowing all aspects of one's self (inner and outer, feelings and other people) to be and to grow in whatever richness is in their seed. This open self is not simply the tool or door mat for others. A colleague just startled me by bursting in my door for the hundredth time. It is good for him that I call into question his manners that we might get along together. It sometimes seems to me that God watches how much I measure out to others and how much to myself. It is Christian to deny one's portion at least a bit (to be on the safe side). It is a genius who can be even fairer than that—to allow myself just equality and nothing more or less. The open self is false if it denies itself, for then it is not open to its own needs. While I admire the uniqueness of the tree outside my window, I can also appreciate this stoutish, middle-aged, Dutch self here. The open self rates itself at least the equal of all others. This means equally as good as anyone else, and equally as bad.

The struggling individual has available the guidance of both the outer forces and the inner. A relaxed perception of the inner lets it voice itself, to reveal what is. One's own dreams can be read. They form the latest daily report on one's favorite stock, one's self. Once I had a long drive alone in the dark of night, so I went to work in my portable laboratory—my mind. If I relax, what fantasy comes? It was of a female companion driving with me. Like a professional psychotherapist I set out to learn of her. Your fantasy may lead to seduction in the dark car, but not with my woman. She was thin, sad, dressed in mourning clothes. Poor woman. I learned her sad tale, how she had been mistreated by her spouse until she no longer knew enjoyment in life. As I became more considerate, she cheered up and seemed less in mourning and even less thin. She really would enjoy doing things but she needed to be coaxed out of her sadness. It was many miles before I fully knew that sitting beside me was my own feelings, and I was sorry how I (her spouse) had treated her. The inner speaks of the more fundamental truth. It cannot do otherwise.

Opening outward one can be guided by all others and even by nature. The relaxed perception of nature restores many to sanity. Opening outward there is increased appreciation for others. Worth is seen in their differences and uniqueness. Insofar as others are given life in one's world, there is a lessening of conflict, difference, opposition. What values from the world shall guide me? The rights of others and all things—the golden rule.

The world as one's own representation seems to be a picture of social and psychological truth. It seems to be a picture or image, a useful way of viewing social matters. Advance but a step further and one enters upon an ancient religious truth. It is not just an image. It is a fact. The oneness of all things is known to l'autre moi. It is intuited in varying degrees and times by all persons, it is referred to in all scriptures (i.e. Matt. 25:40, John 17:21, Rom. 12:5), and in it lies the usefulness of kindness to all. Ego looks skeptical at this. Self identity by its own choosing is a very limited realm. Beyond this is all things which views this identity as one of its images!

There is a beautiful moment called satori or enlightenment given to some at odd times. In this moment the One sees Itself create Itself into all things and the mystery of creation is undone. The One does this in the spirit of endless play. It strives as all things, through all time, to come back home where there is changeless peace. Back home it sees that wandering through all creation, it never leaves home. This is the great self that is all things, that speaks an ancient and nowadays language of symbols using all things to represent itself because they are itself. Out of this insight grows a natural religious life that sees no aspect of life alien to itself including all religions, all ceremonies, all the ancient forms of its expression. Relaxed perception, drinking in the world, sees in itself all things going on.

What I have said here puzzles me! Can I really mean all things are my representation? It sounds lovely, but I would also be saying I am more than just a fellow, nearly a god. The one here who frowns on all this is the silly separated ego who is far from being any god. The point is central. The one who questions, doubts, feels separated is the limited ego.

The one who doesn't notice separation is the other one—the Person among persons.

Yet all the moments I have spoken of are moments when I find myself represented in all things. I can go into inner imagery and feeling and find myself beautifully and accurately represented in endless scenes. At night I dream dreams whose meaning comes clear when I insist on finding how each thing in the dream is myself. The other night some crook (in my dream) hurt me by diverting me into many scattered enterprises. On awakening I caught him instantly because I had allowed myself to be scattered lately. That crook is me. The one who is hurt by this is l'autre moi, the other me. In the hypnogogic state one can pick up instantaneous representations of the most subtle thoughts or feelings. The inner appears to be a vast sea of self-representation.

But the same is true outwardly. The mentally ill person disowns his inner aspects and finds them intruding into his world. It is profitable that alternate sides of self become acquainted. Psychotherapy (and living) is the process of wandering over boundaries to enlarge the scope of one's land. Carl Rogers encourages the other self to come forth, gives it heed, lets it live. This autonomous other self is heard in silence, whispers in feelings, comes calling as a neighbor, modestly stands as background to all that vaunted ego does. When ego sleeps it speaks clearer and implies lands beyond the bounds of the limited self-identity. It is restful for us to contemplate how our thoughts are given, history is given, awareness is given, how it is all given.

The mystery of I-Thou can be crossed in love and friendship when two become one. In appreciating the beauty of the world one savors the things of the world and finds in them one's own inner riches. This world, the world I know, is the world colored with my meanings.

All this becomes more reasonable if the world is no longer viewed as the physicist's abstract, objective world—a totally impersonal other-than-one's self. That world is a conceptual construction convenient to physics but grossly inaccurate in the psychology of persons. The personal world, the only one each of us really knows, is the world painted in the tones of all one's own personal meanings. The world shuts off when I sleep. Its time slows down when I am bored and accelerates when I am involved. My boss is not an objective other. *My* boss, seen by me, known by me, reflects as much of myself as of him. The world of persons is a *personal* world.

Lightning and thunder are beautiful to me. Are they something else to you? Where is the objective impersonal lightning and thunder? They are part of the "reported events" which don't mean much to a person. The impersonal objective world is the one no one cares about! In a deep sense one can know the quality of a person by the world he manages to find. The one who finds a world of dog eat dog, himself eats dogs and is eaten by them. The very good person manages to find good in people. The world of persons is as plastic and varied as people themselves. What has been described here is the world of persons, the one we live in. In it what I see in you is an image of myself. And what you see in me is an image of you. It is a world of mirrors. It is a useless and obscuring fiction that there is *a* world. That is a fiction—a conception of some people that conflicts with their own experience. There are many interlocked worlds.

It is awesome and responsible to consider one's world as one's own representation. Then I am responsible when this world gets mean and small, angry or guilty. I am not only my brother's keeper. I am my brother.

There are times of mergence when the rustle of leaves, or the plaintive call of a bird is the whole of the world. This whole is peaceful, nothing to be done—no questions—no search—all in a simple sound. In rocks or furniture I see stolid, permanent satisfaction with things just as they are. In silent matter is our deepest understanding.

The natural depth of man is the whole of creation. Look at the scope of yourself.

References

[1] Rogers, C. and Dymond, R. *Psychotherapy and Personality Change.* Chicago, University of Chicago, 1954, p. 418.

[2] Van Dusen, W. "The Phenomenology of a Schizophrenic Existence," *Journal of Individual Psychology,* 1961, 17, pp. 80-92.

[3] Oswald, I. *Sleeping and Waking.* Amsterdam, Elsevier, 1962.

[4] Silberer, H. "Report on a Method of Eliciting and Observing Certain Symbolic Hallucination Phenomena." In D. Rapaport, *Organization and Pathology of Thought.* New York, Columbia University, 1951.

[5] Swedenborg, E. *Arcana Coelestia.* London, Swedenborg Society, 1916, 12 vols.

[6] Swedenborg, E. *The Spiritual Diary.* London, Speirs, 1902, 5 vols.

[7] Marcel, G. *The Philosophy of Existence.* London, Harvill, 1948.

XIII

> The future would mean that there is something that is *going to happen*, a "fate." The minute you "see" a future you would necessarily change it by reacting, so it wouldn't be the same future.
>
> Can you work towards the future? It seems to me false to channel your life. You in reality work in any direction from 0°, which is where you are now.
>
> But "now" doesn't exist. You are walking backward into a hurricane and can only tell which way to go by the debris that flies past you. Or you are blown about in this hurricane, judging only the past whether you look into the wind or with it. If you "see it coming" you have already reduced it to a judgment of the past—since all judgments are necessarily of *past* experience.
>
> Best not to worry about direction at all. Just be. Direction is already there.
>
> *from a term paper written by Jim Fiddes, a junior college sophomore.*

What do *I* mean by "experience"?

I look the word up in the dictionary, to help me find my way. It says—*experience:* the actual living through an event or events. . . .

What do *I* mean by "actual"?

It means to me *what happens*. Not what I think has happened or think may happen or think is happening. They are all in the same field as what I think *should* (or should not) happen, which has no reality except as an illusion. Illusions are a kind of reality, but they are slippery stuff to live with. I am better off without them.

In 1944, my husband almost convinced me that I was insane. I had read about this happening in novels and thought that it could happen only to a very weak-minded person, never to me. But there it was. Ninety-nine percent of me believed that I was insane, and I wasn't sure whether the 1% was sanity or determination. I'm not sure now that there's much difference, but that was the way it seemed to me then.

This was in wartime Hawaii, on Oahu. I went to visit a friend on Maui because it seemed to me that I had to get out of what I was in before I could see it properly and begin to make any sense about it. About a week after I arrived there, a man telephoned and asked if he could come to see me—Doctor A, now Major A. I had met him only once, in peacetime. He was a close friend of a New York doctor who was a very good friend of my husband's. How he knew that I was on Maui, I didn't know and didn't ask. My fear stopped me. I thought that my husband must have sent him to check up on me, and that if he agreed with my husband I might be sent to a mental hospital. . . . I have just now remembered that I feared his answer might be evasive, which would shake me still more.

I felt chilly all over, but I said yes, he could come. There was a pause, and then he said he would like to bring with him a Doctor X. I felt cold. I thought, "He's bringing a psychiatrist, too."

In a week on Maui with friends who didn't seem to find me odd, my fears had begun to subside a little the way that a sea subsides after a storm: they were still there, but not so turbulent. Now they began churning around again, kicking up another storm. I had the feeling of falling apart, disintegrating. I was afraid of this then, and held myself together tightly. I said yes, he could bring his friend.

After I had hung up the receiver, I wondered how you can convince two doctors, one of them a psychiatrist, that you're not crazy. I wanted to walk up to them when they came and say "I'm *not* crazy, you know," but I was afraid that might convince them that I was. I thought of all sorts of ways that I might behave to convince them of my sanity, but all of them seemed very unconvincing to me so I threw them out. There didn't seem to be anything to do but be as I would be if these two men were not coming to look into my sanity, but just coming to see me.

The evening seemed to go all right, but after they had gone I began wondering about the verdict. What did they think of me? When the major phoned saying that they would like to come again, I thought they were coming to get further evidence of my insanity. I'm not sure whether they came a third time, or if I just kept thinking about their coming again so there is some feeling that they did.

I had no one to talk to about this, just as I had had no one to talk to about my husband and me, earlier. When I tried to, people cut me off, shut me up. Anything I might say about why they did this is speculation, but *it happened.* And everything went round and round in my own head.

Two months later I was back in Oahu, and the major had been transferred to Oahu too. He invited my son and me to the beach and we went. This happened several times before it was clear to me that he was just a lonesome soldier wanting company, not otherwise interested in me and in no way concerned about me. It was then that I could ask, "What is Dr. X's specialty?" The major said, "Oh, he's a dentist."

The *facts* were few and simple. A man I had met once had called me and asked me to come to see me, and then asked if he could bring a friend. They came. That was the actuality, what I should have lived with—which in fact I *did* live with. But there was *also* all this illusion going on in my head, which I was living with *too*—my thoughts *about* it. I wasn't living simply with things as they are. If the reality had later shown more than that, *then* was the time to interact with it, come to terms with it, moving in any direction from 0°.

If I had persuaded myself that these men were *not* looking into my sanity, that they were simply coming to see me, that would have been illusion too. It might have proved to be otherwise. And no matter how confident I made myself by my assurances, the ground under me would still have been quaking.

When I remove myself from both illusions, then I live with the actuality alone, and when the actuality changes, I change with it, responding to things as they are. What *actually* happens is never so awful as what I think about it. Where does this thinking come from? I have heard of a study which showed that little girls who have been raped are traumatized according to what *their parents* think about it. The more terrible it seems to the parents, the more disturbed the child is. On the other hand, among Hawaiians when I lived with them there were no children in difficulties because of "broken homes" because the Hawaiian people thought nothing of it. The custom of child-swapping by affinity hadn't got entirely washed out, either. When a visiting adult said "I like this child, I will take him home with me," this meant "to live with me," and this seems to have been agreeable all around. Why not? There was a mutual attraction between *this* child and *this* adult. The Hawaiians didn't try to "hold a family together" because all Hawaiians were one family. This made an entirely different social situation, and it could be said that you can't change a whole society. We say that of ourselves, but at the same time we impose our concepts on others and make their societies change. What would happen if we removed these concepts from ourselves?

Is it kindness and "sympathy" to be distressed about others? Or is it as unkind, in its effect on others, as to be cold?

It seems to me necessary that we get beyond *both* and arrive at caring without disturbance. I can remember doing this quite often when I was young—it just happened. But I was told that I *should be* distressed. Sometimes this was non-verbal but I got the message just the same. I was made to feel unperceptive, unsympathetic, a barbarian, lacking in decent human feeling. I *should be* distressed. So then I made myself distressed—and came to see it their way, not my own. Then some others said that I should *not* be distressed—"Don't be emotional!"—and sometimes these contradictory commands came to me from the same person, at different times. I was supposed to be whatever the other person was, in relation to any particular set of circumstances. But I didn't realize this then, and it was very confusing to me that when I did what I had been told to do last time, I got rebuked for it and was told that I "should be" the other way.

"Best not to worry about direction at all. Just be."

The more that I followed my own direction in other ways, the more my direction in this way happened of itself. "Direction is already there." Then, my caring was present, and my not-caring was present too. It was neither and both—just as a blue light washes out both a yellow light and itself and makes a white light, and yellow and blue paint mixed together become a green which contains both but *is* neither. But *I* cannot mix my caring and not-caring. I have to throw out both, and what remains is what I want, the reality of me. No illusions.

When I feel smashed or trapped or bound or pushed, it is helpful to me to ask myself, "What illusion is fouling me up now?" The illusion is what I think about something. The facts just *are*. I know this most clearly through reflecting on what happens in emergencies when "there is no time to think." If I think about it I'm lost—either I am paralyzed and can do nothing or else I move and do the wrong thing. But if I respond to what *is* without thought, then all my past experience relevant to this circumstance is mobilized in me and I act with beautiful precision. *I do not mobilize it,* and "I act" is the objective view—as seen by others, or by myself seeing me after the fact, or after the reality. That is all it means: I am seen as an object. Subjectively, there is no "I" and everything is just happening. This is *experience* to me. As soon as "I" returns there is *thinking about*—or when thinking about returns, there is "I"—and experiencing is lost.

It is sometimes worthwhile for me to *think about,* but when I am *aware* that it does shut out the experiencing which is life itself—gentle love, interest, and many happenings—then I become more discriminating. I ask myself if this thinking-about is before the fact or after. If it is before, then it is fantasy—like my thinking about Doctors A and X. If it is after, then I ask myself if it serves a useful purpose. Usually I discover that it serves only ego, assuring me of the cleverness of "I." There isn't a darned thing I can *do* with it except use it to impress others and myself. This becomes ridiculous. Is it *life* to go about impressing others and myself?

Van Dusen writes, "It may come as a surprise to ego, but in this imagery ego is servant and the inner is the master." With ego = I, it seems to me necessary to be clear which I is me. Is "the inner" *less* myself? Or *more?* When I picked up servicemen in spite of warnings not to do so, I felt that "someone else" had put my foot on the brake to stop for them. But that "someone else" was very much *me*—in opposition to the value that others were emphasizing, which I felt that I should be practical and accept. At that time, this was my "inner" because I had the "outer" too. But when I let the inner act through me, then it became my outer too and I matched. No split. No separation. No conflict. I was whole. Ego-I/intellect stopped the car and then drove it on in *this* world, but the knowing of *that* world dictated when to stop, and when to drive on. Ego is just fine when it acts as my chauffeur.

It seems to me that this half-I/ego/intellect is my small mind, or the small part of my mind, behind my forehead—which I think of as my "intellect" because *I can do things with it.* Real-I simply *is* and *does*—acts through me. When this happens, I have no choice because there *is* only one thing (at this moment) that I want to do. No decisions. I feel then that I am living within a snowflake pattern, myself moving through the center, all things in place around me in a beautiful design. No collisions—within me or with the outside world. I feel like a smoothly spinning top which touches an obstacle lightly and glides away, without damage to either the top or the obstacle.

It feels good to have some understanding of myself, of what goes on in me.

I can get into an argument (conflict) within my intellect, but this is really between other people entirely. It has nothing to do with me. I can also get into an argument between my intellect (other people in me) and *me*. When I accept the *me*, then intellect falls back into its proper place, carrying out *my* wishes which are identical with those of the spirit in all of us.

I think it is the proper function of my intellect to be practical, to enable me to get around and do things in the service of myself and all of us. The error comes in when my intellect makes the first decision, which it can only do "practically" in the service of *it*self and what it sees as *its* possessions. The original meaning of "sin" is to "miss the mark" and it seems to me that this is the way in which I miss it. I have the feeling of hitting the mark when I move through the center of the snowflake pattern—the bull's eye.

When my intellect acts in the service of *my*self (which includes others), rather than *it*self, everything comes out well—with ups and downs but still the total is good in every respect. I do not wish that I had done other than I have. No regrets. Not even a sneaky little one, at any time. Intellect or false-ego has not interfered with *me*. My troubles then are take-able—acceptable, not more than I can bear. In fact, without clinging to them, it feels good to go through them, and as though without them something would be missing—like the seasons, which occur even in Hawaii.

I have always had the wish to be submissive, which doesn't mean that I *was*. If I had been, I wouldn't have wished for it. But when I *made myself* submissive I got into trouble—a crippling of myself—and when I was *not* submissive I often felt quite good about it—clear and strong—and everything worked out well. This was another confusion. Now, the submission that I so much wanted is clear to me. I must not let intellect-ego lead me astray: I must be submissive to the deepest inner and let it spread through to the outer. When I "made myself" submissive without this knowing, it was *to others*, to their demands of me. "I" *made myself* do this. "I" was still controlling me. When I listen to myself, or to the inner I in me, and let this act through me, then ego-I/intellect has surrendered, has *given up* control. The more that I let the right submission happen, the more the inner me seems to *be me*. It is a very difficult transition—like changing barrels going over Niagara Falls—and after all the chaos and labor, when I have "come home," I find—after initial surprise—that this is where I have always been. My life has always contained this knowing but my intellect-ego was not aware of it.

When this happens, I am "all of a piece" (no conflict) which is a very good way to be. "Good" because it has all the qualities of "good"—peace, love, joy and steadiness. This is not present when I "feel good" in triumph (hubris) and this is followed by a "let down" feeling even when it hasn't yet run into nemesis as it eventually does. I learn easily when I live with my deepest inner because there is nothing to block my learning whatever interests me—and many things do interest me because I am not blocked in that, either. I have a feeling of flowing, changing, growing, and am aware of the spirit which moves me—the non-possessive (non-attached) love that the Greeks called *agape* and the Romans called *caritas*. I asked a philologist if he knew an equivalent word in another language and he replied "A synonym in a living language I do not know." To me this love is life itself. How *living* are the "living" languages which do not have a word for it? Do we see only one kind of love because we have only one word "love"? Or have we dropped out the other words because we see only one kind of love, which is possessive? Have we become so obsessed by possession (land, houses, people, things, always buying all of them in one way or another) that this is the only kind of love we know?

These seem to me useless questions. My intellect can become enchanted by its cleverness in pursuing *any* question, either abstract or about my neighbors, but there are some questions which can never be anything but speculation. They can't arrive at knowledge. What I can *know*, by looking into me and questioning myself, is my own love of various kinds, and which kind of love endures in me without change, with everlasting happiness, which is not to be sought in any one or any thing but in myself, although I may experience it with others. What releases me to this is good. I choose it. "The sense of stillness, endlessness: the sense of the sacred and holy . . . for in this peace there are no queries and nothing to be done. It is its own doing." The cessation of desire, not because it has been repressed or suppressed—given up—but because of its fulfillment. Within myself I find my peace. Then, it flows outward, let into the world through me.

Through living this sometimes, experiencing it, I know that "cessation of desire" does not mean the end of delight, but is delight itself. This knowing often slips into the background, sometimes gets lost behind a cloud, but still it's *there* and this softens my desires, makes me more tentative in my wishes and more ready to give them up or let them change their form. It is not a dead end, with things remaining as they are. I am vividly aware of movement and change, and my own participation and involvement. It does not mean that "This is all there is, so shut up and drink your beer." That is the view of life that my schizophrenic friend is afraid he would be forced to accept in a mental hospital, and it is what kept me from going to a psychiatrist when I felt that I desperately needed help in finding my way. It is also what gave me so much trouble in childhood and adolescence: the battle between "grownups" and myself. It is my difficulty with very many people now. They call "supernatural" ("Stay away from it!") what to me is natural, as much a part of me as blood and bones. It seems to me that what is called "supernatural" is dangerous only when ego latches onto it for the enchantment of itself, and "I" become God—above all others—or have some revelations made only to me, my personal self, which I must convey to others who cannot arrive at it without me. Or else I may sit alone in my specialness, requiring that everything be done for me. Separation—not unity. When everyone else is in heaven, I shall be there too.

I have a letter from a young woman saying "I must lead two lives, one to get on in the world of every day and the other one to be really me, to be with the comfortable ones, like you, like, well I can't name them now, but with these people I could be in the really good world for me. But I must learn to live also in this safe conformed world because I have to work and bring up a child and be married to a man I confuse. I must try and confuse him less. I am too odd, too exhausting for others. But when I am with you or with Bev or when I am with Barbara or the mountains—then I am really happy and feel really good."

Odd. Non-conventional. Unpredictable. But the creative *is* unpredictable. Perhaps even God is a little surprised.

Some kinds of therapy seek to make man more predictable. Behaviorism and B. F. Skinner do this too. Certainly this can be done. Parents and society have been doing it for centuries, with enough success to encourage further moves in this direction—and with sufficient failure to permit some changes to take place. Modern techniques make it possible to improve the score for those who wish to control others. There is no question that this can be done, but as one young man says, "A bullet in the head will make a man quite predictable too." The question is: Do I want this—not just for others, but for *myself?*

I don't.

Maurice Termerlin states simply and clearly the other view of man, which appeals to me:

> Unlike scientific goals, the goals of a humanistic psychotherapy are neither predictability or control. In fact, the more successful psychotherapy is, the less predictable the individual becomes, because his rigidity is reduced and his spontaneity and creativity are increased.*

* From "On Choice and Responsibility in a Humanistic Psychotherapy," by Maurice K. Termerlin. *J. Humanistic Psychology*, 1963, 3 (1) p. 37.

That's what I like.

"Unpredictable" scares some people. They think of rape, arson, armed robbery and murder. But the people who make these pursuits their profession are very rigid and predictable. The repetition of their acts in the same pattern is very helpful to those who seek to capture them. They do the same thing in the same way, repeatedly. They can be counted on.

I choose creativity. By that I don't mean "the arts" although the arts may be included. Creativity is not *what* is done but *how* one does it. I can paint a picture or write a book or cook dinner or pick walnuts or deliver mail or fix pipes intellectually—or I can do any of these same things creatively, with my "total organism" in Carl Rogers' words. The cry of many young people is "I want to use *all* of me!" I heard a plumber say musingly, with humility, "I think there must be something wrong with our education. When I am fitting two pipes together, it goes much better when I do it with *all* of me."

Van Dusen writes of "The region that borders sleep." That's part of me too, a part that is very precious to me because to be more aware of it helps me to reach my peace. I don't know whether *people* "vary in their power to even note the hypnogogic experience" or whether this power is in people and shows itself according to the way we live. *In me* the power varies according to circumstances outside me and within me. If I become too tired (I know many people who do not go to bed until they cannot stay awake any longer), then I fall asleep immediately. If I am awakened by an alarm, or wake up with my head full of things to be done, I am awake immediately. If I go to bed with thoughts running on in my head, intellect trying to imagine the outcome or resolve problems without help from the rest of me, then I do not go to sleep until I feel worn out, and then I fall asleep so quickly that I have no awareness of its happening. The unquiet of others or their demands affect me too.

My hypnogogic experiences vary according to my activities, too. When I have been painting, my "visions" are pictorial in vivid colors. There may be landscapes or designs, swiftly changing, or there may be animal people expressing something that I want to understand or know. But everything speaks to me in colors, without words. When I have been writing a good deal, I wake up with nothing but words in my head, and everything is black, white and grey.

Some people say that they "can't" do any of these things. I think by that they mean that they *don't* do them. When a partly color-blind young man was annoyed because his best friend dreamed in color and he didn't, he set himself to achieve this. I was living with him at the time, and for weeks he reported failure each morning. But then one day he told happily of his success. I asked him if the colors were vivid and distinct and he said, "Oh yes! The yellow was as yellow as yellow and the brown was as brown as brown!"

Like everything else that I value, I must make time for it. I must notice myself, and when it is time for me to go to bed. Sometimes I have to make myself quiet before the time when I lie down. And in the morning, I must know that even the slightest movement of mind or body breaks up my awareness. I have to live with my half-sleep (half-waking) gently, and let myself simply notice what is going on in me. When I do this, if I am dreaming at the time of waking my dream continues and I can, while awake, still dream. When I do not do this, what I am dreaming when I awaken may never make sense because it has been interrupted. Not reaching the end may present the *reverse* of what the dream is saying. I once dreamed that a long scroll was moving before my eyes, as though it were being rolled up at the top and unrolled at the bottom, so that I could only see the part before me now. On the scroll were written all sorts of charges (accusations) made against me, and to some extent accepted by myself. I felt more and more guilty as they reeled off, tighter and tighter and more and more miserable: I was the most worthless creature on earth. When I awakened, with my eyes still closed I stayed with the dream, watching the scroll roll on, revealing still more accusations. The last one was: *"And furthermore:* you are a sloppy housekeeper." Suddenly all the other accusations fell into their proper place: they were, in the same way, absurd. What does my housekeeping have to do with anyone but me and the people who live with me? Why should anyone else even waste life having an opinion about it? What makes it still more absurd is that in addition to the variability of other people and how *they see* my "house" (I have been accused of being "too fussy" as well as "sloppy") there is the variability of ourselves, of the quarters we live in, and of our activities. *When* other people see my house affects their view of it (and me) as well as *who* sees it—or: *who* sees it *when*. Only *we*, the permanent residents, can see it "whole." And I am the only person who can know *me* in my entirety.

If I had not let this dream roll on, I would have lived with the feeling of guilt which I had when I awakened, instead of arriving at what the dream was telling me.

"Often the hypnogogic experiences are just out of reach of what can be remembered unless they are immediately recorded, which action tends to disrupt the state." When I was sick for so long and living alone, I learned to "stay with" what was going on until I felt that I had it clear and firm, so that it seemed that it couldn't possibly escape me. But when I made a movement it was gone—utterly. Even shifting a leg or moving an arm or thinking a thought would break it up. Surely this is not completely alien to anyone, when we all have known arriving at something in our heads and then some movement or other interruption broke it up and we lost it?

But there was something else that annoyed me even more, in terms of wishing to present proof of what can go on in me that is intelligent beyond my usual intelligence: when I did manage to move my body without losing the knowing and wrote some notes—clues to the much *more*, which I hoped would serve as an anchor so that I could recapture the rest—I lost the rest. The clues by themselves read like nonsense—as a list of errands in one's private code does to someone else. But my clues were not to familiar *things* which could easily be related to each other by myself: they were clues to ideas, to explanations, to knowings and understandings so great that when I looked at my clues I thought, "It's like trying to put an elephant through a funnel. What comes through doesn't look much like an elephant." Is this difficulty any more "mystic" or "irrational" than what Michael Balint writes about the difficulty that is common to all descriptions of psychotherapy?

> The mind is multidimensional to an impossible degree, whereas any description is limited to one dimension. Language can describe only one sequence of events at a time; if several occur simultaneously, language has to jump to and fro among the parallel lines, creating diffi-

culties, if not confusion, for the listener. A further, almost insurmountable, complication is caused by the fact that mental events not only take place simultaneously along parallel lines, but influence each other profoundly.*

It seems to me that this is why reliance on *verbal* communication gets us into so much trouble. I can never say to you completely what is going on in me. When I use words, this sounds (and is) inadequate and unconvincing. If I then go on to express other parts of what is going on in me, it takes so long, and you become tired of listening. You want to know what is *really* going on in me, as if there were only one thing which could be real. And I, aware of what *was* going on in me, now notice that this has changed and throw in the towel. It is when my total organism receives from you, through non-verbal communication or some mixture of non-verbal and verbal, that I am closest to knowing what goes on in you.

When I was sick and spent so much time noticing what was going on in me, in both my body and my mind (could I be I without either one?), and tried to write about it, I struggled much too hard before I gave up. Intent on my goal, I was trying to force it, instead of considering whether this goal was possible. At last it was forced on me that the only way I could do it would be to write each aspect on a separate transparent sheet, lay all the sheets on top of each other and hold them against the light so people could look through them. Then all they'd see was muddle. And before the first sheet had been written, something in it and all the others would have changed. To have any fixed ideas about myself is nonsense: I can only notice what is going on in me *now*.

That my vast knowings cannot be consciously remembered does not make them valueless to me. Having been in touch with them, I, my life, have been affected. Without them, the little things become so big, they're all there is, there's nothing more. I become a slave to what I "do" and what I "know," when in fact what I do is as important as sweeping out the kitchen (no more, no less), and my "knowledge" is as inconsequential as knowing the capitol of Illinois and how to spell

* From *The Doctor, His Patient and The Illness*, by Michael Balint. International Universities Press, New York, 1957, p. 172.

erysipelas, which do absolutely nothing to enhance my life. They are useful to me only if I need to use them, and then I can look them up. I had to look up erysipelas now in any case, because I have never had ocasion to use it in the 52 years since I won a spelling contest with it. That was its only use to me. I remember a paper by Carl Rogers in which he said the only thing he could remember from one college course was a professor who said, "Don't be a damned ammunition wagon. Be a rifle!"

Any connection between grades and *learning* is illusion. There is not only what I "learn" for exams and forget, but also what I don't learn and get good grades in anyway. I had forgotten about that until last week when I was reading John Holt's book *How Children Fail*. He describes some of the tricks children use to make a teacher think they've learned what they haven't. I remembered then how I hated English grammar and was *not* going to learn about adjectives and adverbs, participles, predicate nominatives, how to diagram sentences, and all the rest. My intelligence was used to make the teacher think that I had learned what I hadn't, so I'd get good grades, which I did. I still don't know anything about grammar. But I can *use* it. When I violate the rules, this is done to achieve an effect that I want. That's the crazy thing about it. I really *don't* know the rules, and yet I know when I have violated them. This is the way it happens when *learning takes place in me.* I know the rules without knowing them. Most of the time I make use of them because they help me to express what I want to say, but I can violate them any time I please, to convey what otherwise cannot be conveyed as well.

The *vast* knowings which come to me never leave me. They are somehow active in my life even though I cannot say (yet) how this activity takes place. Just as I can enjoy the feeling of time with time around it, or space with space around it (nothing crowding me) so I can be aware of knowing with knowing around it and feel more free. Without this, I feel confined within what I know, and that is like feeling confined within an hour or whatever other time I think I have, to get something done in. Then, all other time does not exist for me.

In addition to keeping me in my place as one small me and at the same time a vital part of something so great that "very" would diminish it, it gives me a sense of human potentiality which places us closer to the caveman than to what man could be.

In Carl Rogers' book *On Becoming a Person* there is a paper "A Process Conception of Psychotherapy." In this he describes seven stages which he has noticed people going through during therapy when it is successful. As I read them, it seemed to me that I had gone through many of those stages simply in the course of growing—natural process, which had not got blocked. The last sentence in the summary says "He (the client) has changed, but what seems most significant, he has become an integrated process of changingness." I liked this, but it never had the same impact on me as the last statement in his "Scale of Process in Psychotherapy"—a mimeographed paper designed for professional use. Each stage includes "Examples of Process," and these were interesting, but when I read this section for Stage 7, I was released to joy.

> In time we hope to be able to present examples for Stage 7. It is more likely that such examples will come from the expression of outstandingly self-actualizing people since only rarely in psychotherapy is a Stage 7 level found. For some sense of the qualities of Stage 7 activity the rater is urged to thoroughly familiarize himself with the Stage 6 example from which he can then extrapolate somewhat further.

My thought was (happily) "Stage eight? Nine? Ten? Thirteen? Twenty?. . . ." There is no knowing what man may in the future become—or even now. But I don't think that a "Skinner box" is going to help us get there.

I can look at the present state of man with joy for the possibilities or with sadness for the actuality. I can *choose.* But neither of those choices makes sense to me because I *can* choose one or the other—and slip from one to the other unless I hold myself to my choice. This makes each unreal to me, a fiction. This happened too when I was sick and set out to determine whether my illness was psychosomatic or somato-psychic: I found that I could *decide* for one or the other, but when I removed my decision, there were both. I feel *real* (and steady, solid, firm) when I look the knowing straight in the face, accepting both: "That's the way it is." I feel this way with individual people too, when I accept both their actuality and their potentiality, neither binding them to the one, which is to belittle them, nor thinking of them *as* the other, which is illusion. I feel real when I look at myself in this way too. With the acceptance of *both* realities they become one: my actuality softens and dissolves into the stream of my potentiality and this flowing is me. Where was I? I no longer remember. Where am I? What does it matter? It is already gone. Where am I going? I don't know but I like it.

One of the ways in which I loused myself up by looking outside instead of in was: I was more free than my friends. I kept an eye on the distance between us, and as long as there was this same distance I thought that I was doing all right. I didn't notice that my friends were narrowing down, and that when the distance between us remained the same, I was narrowing down as much as they were. My "superiority" was illusion. But this was a strange period in my life anyway, because even if I had remained the same, that would not have been *growing*. I had settled for a notion of "maturity" which was an end. I didn't notice any of this until I had got so pinched, unfree and unhappy (in spite of many happinesses) that I became desperate. Then, I realized that what I should have kept track of was *myself*—my own direction.

That words and sounds arise around feelings and are linked with visual images is illustrated in a simple way by a trick that gets passed around. A person is shown two drawings something like this—the exact form is unimportant:

He is told that one is a takiti and the other a golooma (or any similar sets of sounds) and he is asked which drawing is which. The angular one is picked as the takiti, the curving one as the golooma. This is so consistent that if anyone does turn up who reads them the other way around, it is likely that he either knew or recognized what was going on and decided to louse things up, or perhaps he so lacked confidence in himself that he gave the answers which were the opposite to his own response.

We speak of colors as loud, harsh, soft, or mellow without even noticing that we do this, and of people as being cold or warm or bright or dull or sweet or sour or crusty. It isn't at all confusing to the senses to be "mixed up" in this way unless *I* get mixed up about it. Our categories in any case are not as clearly defined as we like to think them. In the garden the other day a young woman sniffed and asked, "What smells like beer?" It was the roses. Perhaps not *all* roses smell like beer, but *these* roses did—I had written "do" but changed it, as I do not know even if the roses on this bush always smell the same. I haven't noticed.

That taste and smell are related is something that we are taught, but we can notice for ourselves that other senses are related too. "I smell purple" is not a ridiculous statement but opens me to possibilities which I have been ignoring. It's no more "silly" than "I feel blue." When I was sick, I noticed that pains "have" color, shape, movement, temperature and voice. It seems to me that the ability to notice may be more a matter of our noticing than of individual differences. If I "forget my pain" (as I am told to do) I cannot notice it. And so with dreams. And with the experiencing of something much greater than myself in me. Whatever I dismiss, I cannot explore. Like the doctor in New Jersey who was furious when penicillin was discovered because his bacteria cultures were killed by molds too, and he kept throwing them out instead of studying them as Alexander Fleming did. What else are we not noticing that would be helpful to us?

Musical sounds can also be descriptive. A. H. Maslow tells of a party at his home at which Max Wertheimer "played a few phrases on the piano to describe each person present. He was very good at this because everyone always knew immediately whom his piano was referring to. For instance, I was able to guess successfully whom he meant in all but one case and then I noticed that everybody in the room was looking at me. I felt that there was certainly a moral here about self-knowledge. This was characteristic, as I remember it, that the person depicted could not tell from the music that he was intended as the object."

Is there a "moral" here, or is it simple mechanics? When I am looking outside, I can't notice me. When I ask people to name the planets, it is very unusual for them to include earth. When I have been asked to count the people present, to set places for dinner, I have often forgotten to include myself. When I have not this intentness outside, it is much easier for me to include myself. It seems to me that my intellect does the "seeing" when I exclude me. When my total organism is receiving, *I* am included—and then I also know that whatever goes on around me is to some extent affected by the fact of my being present. Literally, nothing is the same without me. Even the thoughts in my mind have an effect both on my seeing and on what is being seen, in terms of people. I think it was James Jeans who said "Man cannot lift a little finger without disturbing all the stars." It takes a thought to lift my finger. So even my "unconscious" gets into it, because my finger may be lifted as a response without conscious thought.

Awareness of these possibilities gives me pause, makes me more tentative and gentle, and at the same time more quietly insistent that I must have what is good for me, for all of us. It is best for me (for all of us) to leave what makes me forceful or explosive. "Controlling" my force, in the sense of hiding it, does not remove it. If I cannot let it wash away, dissolve, so that my gentleness is present, then it is best that I remove myself (not necessarily forever)—however much it may go against the conventions of the society in which I live. Among some people labeled "schizophrenic" and among Hawaiians, Hopis and Navajos, I have noticed often this withdrawal *before* anger or explosiveness takes place. "If I stay, I shall say what I do not wish to say." "If I stay, there will be trouble. I wish to leave before that happens." When I was young, several times it happened that I was "in love" and enjoying serenity and peace. Then I noticed frictions building, and I left before they could be damaging to either of us. This seemed to me best. I did not wish to be harmful to the other or to myself. I was called "callous." It did not seem that way to me. But when I had been called "callous" often enough, I thought I was, and stayed—with damage to all of us, including the children, which it took years to heal, and some of it is not entirely cleared up yet. The "loose morals" and "irresponsibility" (about work, with employers) of Hawaiians, Navajos and Hopis seem to me to make better sense, to be *more* human, because they are less damaging to other humans and to my humanity.

That we "criticize our neighbors for what are our own faults" could be easily misunderstood. When my neighbor chains his dog on a short leash and blows smoke in the dog's eyes, I don't like this, but it is not something that I do or even have ever wanted to do. But when I have "held" someone insistently so that I could tell him what he did not wish to hear, is that *different?* It is not the act itself, but what the act is doing.

As a small child, when my mother told lies to someone who had come to the door, I wished she wouldn't do it. I was "criticizing her"—to use grown-up language. But I did not yet know this way to think about it. I didn't blame her. She was the way she was. It felt to me simply that I wished she wouldn't do it because it opened up a path to me that *I* did not wish to go, and if she hadn't done this, I would not have known it was there. I knew only honesty until something else was shown to me. It was a very strong factor in my close identification with my father that he was honest. This I liked. I felt steady. He also made it possible for me to be honest with him—when I was small. Later on, he didn't. If I want to be told the truth, I must be willing to accept it. And if I wish to have love, I must be willing to receive it.

Now, when I feel critical of other people, I often notice that what I am really feeling is "Please don't do that. It is what I don't want to do, and I so easily fall into it when you are doing it." This is certainly true of stereotyped chatter about world affairs, gossip, literature, anecdotes, and such, and people who are forever planning, who never live *now*. Other people's intellect/ego calls out mine, and then I don't like me. I would like there to be an irreligious order where I could go at times, to be with others who wish to work out the same corrections in themselves, to strengthen me in making mine.

However, it can also be simply and directly true that what I criticize in others may be something that I do, too. It is certainly wise to look into myself and check—and also ask others whether it seems to them that I have this fault. It may seem to me that I am only "reminding" someone else (being helpful) while to others it may seem that I am "always nagging." Then I may understand better how someone else's nagging (as it seems to me) appears to them.

Among my friends, I have many who have been labeled "schizophrenic." I feel a very strong bond with them, as I did also with my husband, for whom I never had a label although his mixed-upness was certainly extreme. When we were in deep trouble together one day I said, "Well anyway, *basically* we agree." He said, warm and strong, "Yes. It helps, doesn't it?"

This basic agreement, which I find among all my "schizophrenic" friends and me, is a knowing that all men are brothers. This leads to difficulties in the "practical" world. My husband, as a doctor, could hardly bear to make himself send out bills. He said with agony, "If someone fell down in the street, you'd do anything you could to help them. You wouldn't think of *charging* them." For this, in many quarters, he was berated and ridiculed. There is very much the wishing to be responsive, to do anything one can for others "for nothing." This is true of my very seriously disturbed friend who is not able to do this often, and who finds his own problems "more interesting to me than anyone else's." When he cannot love (respond), he tries so desperately to love that his whole life *is* his trying and his failure.

Another of my friends who has this label, and whose sister has been hospitalized for nearly two decades, has the "practical" difficulty too. He tried buying old cars and fixing them up to sell them, but when a person wanted to buy a car, my friend would ask, "What do you need it for?" and when this had been explained my friend said, "Oh, you don't want *this* car" and told him what kind of car he should look for, which my friend didn't have. He stops when there is a rock on the road. He gets out of his car and moves the rock to a place where it won't endanger people's lives. He goes along with people, unobtrusively, to see that they arrive safely and easily at their destination. They don't even know they have a guardian. He does these things when he "should be hurrying to get somewhere" or "should be taking care of his family."

To him, all people are his family and his caring cannot stop at home.

Their way of living is often very beautiful—the inner me coming through, letting into the world the humanness that we exclude not because we don't like it, don't think it would be a good way to live, but because it isn't "practical." Sometimes their love for all others is seen as a threat to monogamy, although this love has nothing to do with age or sex. As one of my friends has written me, "It isn't limited to the right religion or the right color or the right size or IQ or type of person. It is the love of a child who knows none of these things. It's funny, really, because it does not make any sense at all, that the child should love so much and so many and so often and still be able to love again."

And a funny thing is, in an adult this is often seen as "dangerous." How can there be anything for anyone to be afraid of, either the one from whom love flows or the receivers? Spread around like that, why should there be fear? Because the fear—which I now know as possessiveness—was not in myself, I couldn't understand it. I felt so completely innocent and was charged with being guilty. When I was still strongly myself and clear about what was going on in me, this seemed to me just silly. But as it continued there was more and more pain in me, and more and more confusion. Sometimes I was told that I *couldn't* love as many people as I said I did, that I was "making it up," that it was "a lot of nonsense." And then I began to wonder if I *was* "making it up"—and sometimes I thought this must be true, because so many people said so.

Still later, this was seen as "sex." Naively I said, "Oh, no, because I feel this love for women, too." In me it was clear that this was not "sex" but that was not the way that others saw it. "You'd better watch out!"

And now, when I insist it is not sex, very often people say that I am afraid of sex. It's not that, either.

Not this, not that.

But something *more.*

Yet when I say it's *more,* then I am often accused of thinking that I have something special that is not available to others.

It's not that, either.

But there was a period when I became as confused about

it as anyone else. Now that I'm really clear, I don't need to talk about it—but I still tend to hide my love so that others will not misunderstand. When I do not let it flow through me in its own expression, it dies in me a little, and I die a little too—not just "emotionally" but in my body too. I become more old and less alive.

Fortunately, in recent years, I have been coming together more frequently with people of all ages and both sexes who know their love for what it is. They recognize the love in me. Released in both of us, it comes powerfully together. One of them calls it "the wordless resonance."

I don't know why this is happening now so much more frequently than it did earlier, but I think it may be that the tabu on "mysticism" is being partly lifted. I *know* that it is true that all of these people have had experiences that are called "mystic" although this is something that I have learned about after our getting together, not before, and it is not very much talked about.

One day a doctor telephoned me because he was sick and feeling lonely. I had met him only once before. Over the telephone he spoke of camping in the Sierras with his family. Sitting around a fire at night, with a bear tumbling around in the brush and all the night-sounds fully sensed with all his senses, he was overwhelmed by the truly indescribable, the being a part of nature, of all the universe, with all the knowings that come to a person at such a time. It made most of what we do in our "normal" lives seem nonsense to him, like counting the feathers on a bird or running to catch a train which it would make better sense to miss.

I think that everyone must have this sense of futility at times. I don't mean the one that comes about through failure but the times when I sit among the baggage of Success and wonder "What does all this have to do with me?" A feeling of its unimportance, of the senselessnes of so much that had to be done to arrive at it, and still has to be done to maintain it. A questioning of why it should be maintained at all. When I feel this way, I am usually told that it is because there is something wrong with me—that I "should be happy." There *is* something wrong with me, but it does not make me right to convince me that it does make sense, because it doesn't.

I said to the doctor, "With this knowing first, then everything we did afterward would be different."

"Yes."

I didn't know then that this had already happened to him. He was still a doctor, with the same offices, the same patients, and the same hospital appointments, but he was viewing "medicine" in a different way—in line with Hippocrates' "It is more important to know what kind of a person has a disease than to know what kind of disease a person has." He is also insisting that there be a Department of Caring for People in hospitals and medical schools. This is a very practical idea, but even if it weren't, most of us would appreciate it.

. . . . I have been asked to explain that.

It is supposed to be very bad to say anything "against" the medical profession because then people will lose confidence in doctors, and that in itself will militate against their getting well. But I think that we patients can keep our heads—especially if we see that what we criticize is not just callousness on the part of the medical profession, but a general misunderstanding of what "caring for people" means.

I do not generally feel cared for by doctors or in hospitals. There is little or no response *to me*. I am approved or disapproved according to a stereotype of how a patient *should* behave, and what I say is believed or disbelieved according to a stereotype too. In Michael Balint's work with a group of doctors, he found that each doctor has his own idea of how patients should behave, and that he has almost a salvationist determination to see that they *do*. I found, in moving around and being a patient of a number of different doctors, that each one had his own ideas about how I should behave, and that when I did what they wanted me to they were nice to me and when I didn't, they were annoyed with me. Because of my dependence on doctors at that time, I was afraid that they might kick me out if I displeased them too much. So I tried to be what they wanted me to be, what they demanded of me. During the period of my sickness when I had no money, when I felt dependent in this way too, I made the same effort with regard to others. I became quite neurotic in the process. It seems to me that this is one of the things that gets children all mixed up, so that they become *problems* then or later on. There is the fear of being kicked out, not given the care they need which they are unable to provide for themselves, and so they try to make themselves be and do what is required of them. This louses up anybody. So it is not *simply* "lack of love"—although lack of love certainly enters into it, for "I" am loved only if I let myself be pushed into a pattern which excludes *me*. I am not in any sense respected, and neither am I *cared* for, no matter how much may be done "for" me.

When a medical treatment is harmful to me, I can know this *before* it can be apparent to others, but I am required to go on with it, and am "troublesome" and "unreasonable" if I protest, and "unmanageable" if I refuse. I am subtly or unsubtly punished for my disobedience. I am placed in the position of having to fight, when what my body needs, to put itself in order, is to be at rest.

When I ask questions, I am answered with evasions. These evasions, which are supposed to be "for my own good" and so that I won't worry, tend to make me worry about what it is that is so disturbing that I can't be told about it.

In a hospital, I am supposed to eat all my food. When I do, this is favorably commented on. I am a good girl. When I don't, this is unfavorably commented on. I am a bad girl. When I was in the hospital following the birth of one of my children, one night I had bellyaches—and I could hear all the babies crying. It seemed to me that something in the food I had eaten must not be good, and that in view of my bellyaches it would be best for me not to eat breakfast—or at least not until I had talked with the doctor. I explained this to the girl who brought the food. She went out and a nurse came running in, and I explained it to her. I said that I wouldn't eat until she had called the doctor. Another nurse came in and tried to coax me to eat my breakfast. It seemed to me that I was besieged by nurses insisting that I eat, although probably it did not happen very many times. But I was tired, and that was the way I *felt*. This was not good for me, to feel that way, using up my strength in "fighting" that was really just holding out, when my strength should have been building up in me.

Finally a nurse came in waving her arms and saying "Don't eat your breakfast!" as though she were giving a command which I had been violating. She had talked with the doctor.

In another hospital, I sometimes didn't eat because the food which was brought was not good for me. I didn't know why, but it gave me trouble. I was not supposed to decide this for myself. But it was my stomach which told me. Six or seven weeks later, the doctor knew why and was very apologetic. He had forgotten something which showed up in the x-rays. I couldn't blame him, because he had told me what showed in the x-rays and I had forgotten it too, because at the time we were more concerned about something else. It hadn't become a problem yet. When it did become a problem, I was the first to know because the problem was *in me.*

This is not the place to go into all the other ways that I do not feel respected by doctors and nurses. It can be lumped together in the statement that there is little or no interaction. I do not feel cared for, and when I do not feel cared for, I'm *not*—however much someone else may feel that he is doing "for" me. *I* am not consulted.

This is the same *thing* whether the situation is doctor/patient, teacher/student, parent/child, government/Indians, social worker/welfare recipient, or any other authority/slave situation.

Supposedly things are done *for* me, but in fact they are done *to* me. Because I am done *for* in the minds of the authorities, who genuinely want to help me, I am "ungrateful" if I resist, and they feel hurt when I do not thank them for what they have done "for me."

But I do not feel done "for." I feel left out.

And in fact I am. I haven't been given a choice. I haven't been told what the choices are. These have been decided *for* me.

Not all doctors do this. I can choose whether I want to go on with a doctor by asking his questions and seeing what kind of answers I get, and how he regards *me* when I ask questions. If he respects my questions (and me) and gives me clear, direct answers, I have confidence in him. No matter what criticisms may be true of the medical profession as a whole, I trust and respect *him.* If he is annoyed by my questions and brushes them off or gives evasive answers, then he does not, in my language, care for *me.* He does not accept the fact that these questions are important to me or I would not ask them. He does not understand that clear answers are necessary to my peace of mind and body. In a town where I was a stranger, I once chose a doctor by overhearing a woman say, "I'll never go to Dr. X again. He always says he doesn't know." I never made a better choice. He knows lots more than most doctors, and he really cares for his patients.

He also has a full schedule and a long waiting list, even if one woman and perhaps some others did drop out. He isn't afraid to say "It's my fault" when it is. He lives with reality, including the reality of non-possessive love.

So much for that.

A few hours after talking on the telephone with the doctor who spoke of his mystic experience in the Sierras, I noticed that my own difficulty with the "safe conformed world" seemed to be considerably less. Whenever I have this acceptance by another I feel released to myself. Then there is no pushing "to get things done" but myself moving into them as a part of the doing. Even when the same things are done, they are done differently. There are not "two worlds"—just this one. There is no split in me. The "safe conformed world" continues to be, but I don't really accept it *or* reject it. It's more like getting nimbly around conventions and bureaucracy. I have a feeling of good invisibility, of not really being noticed except by those who are in my world with me. Perhaps this comes about by my not noticing those who might disapprove of me: their disapproval does not enter into my world or me. Perhaps it is that my disapproval of others has been removed with the disapproval of myself? What does it matter? They come and go together.

This is the world I long for when I have lost it—not the womb or my mother's breast, but the living, vital *now*, alive and pulsing. I am not lost in the timeless, nor in time; in unconsciousness or in consciousness. They meet in me. I am exactly where I belong—a point moving in time against a background of eternity.

When I am in this world, non-possessive love is fully present—gently touching, never clutching. I do not have to make corrections in me to arrive at it, as I sometimes do at other times. In this world, this love just *is*. You come toward me. I have the feeling of a smile. You leave me. The same lightly dancing happiness remains. I miss you sometimes. I really do. You come into my thoughts and you are not here. But at the same time I am not missing you, for you are still with me. I wish you well. I don't know what you are doing and I do not need to know. My wish is a feeling without words, "May all be well with you."

You are here. That is good.

You are not here. The goodness remains with me.

You are with someone else? I trust that you are happy.
My life continues, and my life is me.

At this deep level we are together and a distance of ten thousand miles is right next door. No, not next door. Right here. I do not long for you, for you are with me, and this I like. You do not occupy my thoughts, my life, but are with me like my breathing which goes on, keeping life in me. All the many yous I love are together in me without clutter, those physically near me, too. There is responsiveness in everything I do, and the whole world is very much alive.

At such times, sexual intercourse is not important to me. It never has been, at any of the ages of my life, when I live with all creation. I have read in two very different places that the sex rituals of more primitive peoples were for the purpose of arousing sex in people who were too happy to be bothered with it. They weren't attending to their duties.

Now, a very attractive and capable young man tells me bitterly, "The only place you can get close to someone is in bed," and a woman in her forties says, "The only time we are together is in bed and what kind of marriage is that?"

I have a letter written forty years ago by a man who was (in) famous for his championship of sexual freedom, although his "followers" used this in the very unfree way of insisting that everyone *must* engage in it frequently: *the* solution. Although they were at the same time very scornful of panaceas. They are bound by or to sex—like the "non-directive" therapists who were bound by directing themselves to be non-directive. The letter says, "I was *very* happy all evening. In some ways, one gains by keeping passion under—its mental transmutations have a wonderful quality." The wording is quaint because the man who wrote it was not an American. He also wrote within the concepts of his time, so he interpreted his experience in terms of Freudian sublimation. "Keeping passion under" sounds like lack of spontaneity and freedom, but in fact he did his own choosing. It was not convention which dictated his choice, nor was it anti-convention. In terms of his own life, there was something he wished to explore at this time. As so often happens when one explores, he stumbled into something that he had not anticipated.

I switch to my own life when I say that I would not call it "mental transmutation," although I would agree about the "wonderful quality." Through this, I understand something of the religious view of sex which seemed to me at one time so misguided. I still think it is, in the way it is presented. Where it went wrong, it seems to me, is the same mistake that the sex (and other) people make: seeking to *impose* on others what must be voluntary. I have to be *willing*, as the man in the letter was willing. That seems to be the difference between heaven and hell. I am not pushing my view of this on anyone, but I think that it should be offered along with other views, for people to choose for themselves, in an exploratory way. To make it a belief or dogma would get us into the same trouble that we are in now via the opposite door. "You must." "You must not." Conflicting orders.

All that I wish anyone to do is look into himself, his own experience, and do his own questioning. "And I would fain have you remember me as a beginning" not "the end."

To me there is not only the question of sexual intercourse or not, but also if there *is* this intercourse, *how* is it gone about? I do not mean this in the external sense, to which many books are devoted. To follow them is to go about things back-end-to. The emphasis is on *pleasing*, pleasing the other. Trigant Burrow speaks for me when he writes that

> Love is unity, participation, understanding. It is simple, harmonious, unquestioning. Love is one with life itself. It is life in its subjective relation. Cognition, on the contrary, pertains to contrast, demarcation, distinction. It is close kin to pride. In other words, it is synonymous with acquisition, aim, calculation. Hence it is kin to self-interest, to desire, that is to say, to sexuality (as distinct from the unification and spontaneity of sex).

It seems to me that sex is much better arrived at without words getting between us, so that I may be with that love which is

> . . . without object. It is whole, spontaneous, free. Sexuality has its object, its divisive gratification. Sexuality always clashes with love. It is self, and love is precisely the unawareness of self.*

* From *Preconscious Foundations of Human Experience*, by Trigant Burrow, Basic Books, Inc., New York, 1964. p. 30.

Then—how can I love you or you love me, or I please you or you please me? Love and happiness are simply present.

But it is very easy for me to become trapped in conventions, to play a role instead of being simply myself, with no thoughts about me. Even when I wish to be free of roles, most often the other person insists on it, will not permit my freedom. Occasionally, someone does.

Paul E. Lloyd worked in physics until he went through what is called a "schizophrenic episode"—an extraordinary way of labeling hell—and came through on the other side knowing what he wanted in this world and letting it into the world through himself. He made the Western Behavioral Sciences Institute possible, and he made it possible for this book to be written and put together. His support of me so that the book could be written got us into difficulties for awhile which neither of us wanted: he didn't want to feel like a benefactor; I didn't want to feel like a beneficiary; and yet there were these notions in our heads about it. But because we both didn't want them, we threw them out and got beyond them. Who should be grateful to whom? We both were happy.

This breaking free was very wonderful to both of us and our love was present. Then, conventional notions of man/woman relations tangled us up for awhile, though none of this was for long. Man/woman as (conventional) "friends" was less than what we wanted, with the distance carefully maintained, saying "We are friends, but we shall not let it become more than that," avoiding real closeness and interaction. Man/woman as (conventional) "lovers" was less than we wanted, with its demands of closeness and its possessiveness. Either one would *require* something of us. We threw them both out and arrived at what we wanted—human to human. It cannot be described. It's too ambiguous. There's never any certainty about it. We like it.

XIV

> It would be difficult to conceive of any phase of mental experience less representative of health, growth, and conscious evolution than "normality."
>
> Trigant Burrow*

We are told that we are not practical, and that in itself we don't mind so much, but it is said in a way that means that we are irresponsible and that hurts very much. Our love is called "sex" and sometimes we become confused by what we are told. And our knowing of a better world is called "idealism" or "mystic" or "imagination," and when we would like people to come into our world to know the joy of it, they become frightened, and insist that we must come into theirs. But their world is a world we know and we don't like it, and our world is a world that they will not explore.

We want a world that is warm and friendly and they say "This world is that"—and go on stepping on or over others to "get somewhere." "The trouble with you is, you have no ambition," they tell us. And we feel ashamed and guilty, and crawl into ourselves and hide.

They convey "You're worthless" and, wishing to be good, we feel bad because we are worthless. And if there is enough of this, unleavened by love and understanding, we enter into a kind of paralysis and can do nothing. We are feeling so guilty. We can't do "what is right."

And if we do do these things that are "right" and "good" we have to make ourselves cold to do it, cold and hard, so we won't slip and be human, so that we won't remember ourselves and know the pain. "Forget it!" "Keep your mind

* *Ibid.*, p. 44.

on your work." The important thing is to stay on top, so you'll be respected. But where would be "the top" if there were no bottom, and why should some people be the bottom?

Looking outward, to others, for what I can only find in myself. The less that I have self-respect, the more I need respect from others to take its place, and the respect of others can be lost so very easily. Just one slip. . . . So I must always watch my step, be wary. The world is a jungle, and I have made myself a beast. Or is it making myself a beast that creates the jungle? Or, the world is seen as a jungle and I want no part of it.

"It'll make a man of you." A man? Or does it mean, "it will take out of you what I don't have?" "What I have given up, you must, too."

Is it human to be beastly?

Or is it just human to make a mistake?

"Egotism is plain unhandy. Insofar as I am Somebody, I cut myself off from other-buddies. I am trapped in should/oughts, moment by moment weighed and evaluated in the eyes of others, a victim of opinion (mine and theirs), constrained to put on my best shopkeeper's face and coat." This happens insofar as I am Nobody, too. It is the same whether I am looked up to or looked down on (opinion—mine and theirs). My only freedom is in being equal—not with the Somebodies, not with the Nobodies, but equal with all of us. I cannot feel myself to be above someone else without putting someone else down. I cannot feel myself to be below someone else without putting someone else up. "All men are equal if only we will regard them so."

When I live with equalness, the Nobodies usually like me but the Somebodies are not always so happy about it. I am not showing them due respect. When I was twelve, my mother told me that my first-grade teacher had said, "I don't know what to do with your daughter, Mrs. Fox. She treats me as if she were Lord Mayor of London." I was very sad to hear this, for I had loved and respected Miss Livingstone very much, and felt strongly that she was my equal. I had much to learn from her, but still, she was my equal. I felt that way about Bertrand Russell, too, when I was in my twenties. He liked this. To me, it is democracy. Equalness *is* with honesty, and accepting me as I am. Am *I* any different when I am a university press editor or when I am working as a cook in a boarding house? It is good for me to switch around this way in my living and notice the nonsense—that my views carry more weight as an editor than as a cook, when I am still the same person with the same views. It was good for me to notice years ago that I was deferentially received by clerks when I was wearing a British wool suit, with my hair in place, and ignored by the same clerks when I was wearing Levi's and a flannel shirt, with my hair blown about by riding in a jeep. In both cases, *I* was not being met. And who—or what—was meeting me?

"The open self is false if it denies itself, for then it is not open to its own needs." It would be nice if it weren't so easy to get mixed up. When I came to know my own need not to use force on others or myself, and that I must "let things happen" and all would be well, I was so carefully guarding against imposing on others (looking outward) and so sure that what I needed would come about—and then it didn't. It was a long time before I realized that *I* was making this non-happening happen by not permitting the happening of myself, that I *was* using force on me. We come upon these knowings in ourselves and misconstrue them, in terms of living by them, as I did. "I don't have to do anything—it will all come to me." The words are all right when I read them in the right place in me. I should not do something intentionally *to* produce something. But if I do not let myself happen, *I can't move into it*, which is where the happenings take place. I still make this mistake sometimes, but I'm quicker at noticing, now. When I knew that it was right for me to leave a place, I was waiting for circumstances or people to make this come about—to make it happen. I knew it was right: therefore, it should happen. Then I noticed that I was waiting for the decision to come from outside, *forcing me.* At that point, I made the decision myself. In reality, it had already been made in me but I was not letting it act. When I did, *then* other happenings took place around me and they happened (on the whole) very well, not only for me but also for others—in ways that I could not have foreseen.

True quiet means keeping still when the time has come to keep still, and going forward when the time has come to go forward. In this way, rest and movement are in agreement with the demands of the time, and then there is light in life.*

The razor's edge.
The bull's eye.
A thin line.
A small dot.

So much easier to miss than to hit, to fall on one side or the other. And yet I seem to be so made that I can do it, if only I can get rid of what prevents me.

Then, I "take my time" when my time should be taken, permitting others to take their time too. I do not hurry them or me. As I have already mentioned, when I hurry it takes longer, and my peace is lost. Or as one young woman said, "I feel that I'm going 90 miles an hour inside and 5 miles an hour outside." When I slow down, the inner speed decreases and the outer speed increases until they match. This is my subjective feeling. (When a small child seems to me to be dawdling and protests "But I *am* hurrying," he may be speaking the truth of himself.) Objectively (as I am seen by others) I have slowed down, may even seem "dull," but although I "do" less, I accomplish more. I am in harmony with me, and whatever I do goes easily and well. I do not do all that I would like to do, but I do all that I *can* do (at any time). By limiting myself, I have the feeling of being free. And what is "freedom" but having the feeling of it, throughout my body and my mind, with all the world open to me?

"Time slows down when I am bored and accelerates when I am involved." It seems to me that this is true only when I am in the wrong place, where there are these either/ors. There is another place (or world) in which time changes differently: I am not bored. I am involved, and yet I am surprised at how much has got done, that a usual day's work has taken only a few hours. I look at the clock and am astonished that so little time has passed. During those few hours, time has not seemed to be either fast or slow. Time and I have flowed together, so how can I feel it moving

* From *I Ching: The Book of Changes*, Richard Wilhelm translation, into English by Cary F. Baynes. Pantheon Books, New York, 1950, p. 214.

or note its speed? It has neither got stuck nor run away from me. I can only be consicous of time when I am not living fully in the present.

". . . no longer viewed as the physicist's abstract, objective world—a totally impersonal other-than-one's self."

I wonder. . . .

When I was dying rapidly, I knew two views of my body. With my eyes, I could see a good deal of me in the usual way, but this seemed unreal. What I *lived* with was a different kind of seeing with my mind, in a way that was very puzzling to me—new and strange, and at the same time it seemed to be more real than my other view of me. For two years afterward I looked through books in search of a description of a similar experience, to reassure me, and found nothing. Then I happened on a description in a magazine of the "atomic physicist's two desks"—the one we all know which can be bumped into or sat on, and the other one which is all whirling atoms with space between. This was, with absolute clarity, a description of the two ways that I had "seen" myself: The usual one, and the other like motes in a sunbeam, both of which I lived with for several weeks. When x-rays were taken of my body, this seemed silly. Anyone (it seemed to me) could see there was nothing there, just by holding me up to the light and looking through me. If the doctor had come into the room and asked "Where are you?" it would have made better sense to me than his speaking to me as though he could see me.

I have had the same kind of experience when I wasn't dying. I received a letter from friend M sharply criticizing friend C. Knowing the life of friend C, I felt hurt by this lack of understanding. I wanted to write to M and slash her for slashing C. But at the same time, I knew M's life well and why she did this, and also knew that C would criticize M in the same way. So why should I attack one, to defend the other? And if I did, wouldn't I be doing *the same thing* —doing myself what angered me in my friend?

I *knew* this, but I couldn't accept it. I still wanted to be mean. I felt mean toward M, and this wanted to get out. But that was altogether unreasonable and I fought against it. My attention was not on refraining from writing the letter, but on trying to get this nonsense out of me. This was such an inner struggle that I felt I understood the meaning of "exorcizing the devil." I felt so tied in knots that it seemed I never could get out of them. But then I broke free—and also went into a "mystic" experience of the kind that knows that "I am my brother" (and my brother is I), and everything went into dancing atoms. It's like a sort of shimmering, with no clear boundaries. Sidney Cohen says "Our visual mechanism must have evolved with the goal of keeping the organism viable rather than with the aim of seeing things as they are." Do I sometimes see things more nearly as they are? If I do, is this "hallucination"?

It seemed to me that since *all* was these atoms with space between, everything could pass through everything else—that I could pass through the brick wall outside my window as long as I lived completely with these atoms, and did not let myself become shaken by my other knowing of "brick wall" and "me."

This seems to me as fanciful as it possibly can to anyone else, but it *was* my experience—and not my imagination, which is another kind of experience. I am capable of fantasying—and sometimes have done this deliberately. It isn't the same. I can differentiate between the two, just as I can differentiate between tea and coffee, or between the sound of a tractor and the sound of a bird.

After this experience, I was free of the bind. My recognition of the reality of my two friends was like something seen and accepted by all of me without question, without fuss—like the fact that they both had brown eyes. There was nothing that I need do about it—nothing to be done.

I do not understand this experience of dancing atoms but it *is*, and when I had accepted it, I remembered other experiences of this kind. My selector seems to do this: shut out one experience, and all other experiences of the same kind are locked up, hidden from me. When the lock has been broken, then all similar experiences are released too. (There is another selector which I haven't mentioned. That one is much more discriminating. Or perhaps it just discriminates in a different way. It selects from the archives precisely the past experience or experiences which will tell me what I want to know—and nothing more.)

One of the times when I had the experience of atoms and all oneness was in 1947 when I was staying in a remote

Navajo community where my friend was the only non-Indian resident. She worked there, in as many capacities as she was capable of. One of these was nurse, although she had had no nurse's training. One day she went off to be with a Navajo woman who was bearing her third child, to help in any way that she could. She came back looking shaken, and said that she thought the woman might be going into convulsions after the birth of her baby. She phoned the nearest doctor, seventy-five miles away over very poor dirt road. The doctor said there was nothing to be done but keep the woman warm and wait.

We gathered up all the jars we could find in the house. Then, my friend asked me and another woman visitor to go back with her. The other woman declined. It seemed to me that my friend was in need of moral support, and that I had none to give her. I was shaken myself when I thought of all that I did not know, how unequal I was to anything in this situation, and my complete not-knowing of Navajos as I had only just arrived there. I wanted to say *No!* But how could I desert my friend who needed me? I said yes, and went.

The Navajo woman was lying on a sheepskin on the ground in front of a tent that had absolutely nothing in it. There was a shade made of branches in front of the tent. Her other two children, about three and one year old, were sitting on the ground near her. Her mother sat beside her, sometimes rearranging the juniper twigs around the patient's waist, which were to help her come through all right. The patient's husband and father stood a little way off, with their backs to us.

The two men built a fire and put a washtub of water on it. When the water was hot, we filled the jars and placed them all around the woman, completely encircling her, and put blankets over her and them.

And then we waited.

The Indian capacity to wait is so vast that there is nothing to describe it. It has seemingly no boundaries and no limits. This patience entered me. And then I began to feel the pain that was in all the Navajos, that was not expressed in voice, face, or gesture. I didn't imagine it. It didn't begin in me, as my "sympathy" often does. It came to me, was received by me, and I shared it, tenderly. There was this tenderness everywhere. Once, the younger child began to fuss. The three-year-old put his arm around the younger one, and the fussing stopped.

How long we stayed that way, I do not know. The men kept the fire going and the water hot, and when the water in the jars cooled, we refilled them.

And then at last there was a change in the patient patient. She seemed to smooth out all over. Knowing nothing of what had been going on in her, still, it seemed to have left her, now.

My friend sent the men to bring a bed and some other things from the government school and put them in the tent. She asked the men to move the woman to the bed. Then she got two pans of water and two wash cloths. She handed one of each to me, and started to wash the woman's face. She nodded to me to wash the woman's bloody thighs. I hesitated. It seemed an invasion of the woman's privacy, this woman who could speak no English, to whom I could not express my feeling, and I did not know how she felt about me. But then my thinking stopped, and my humanness came through. Gently I washed her. I looked into her eyes and saw such trust in me to do what I could, not expecting more, and somehow this trust in me, a stranger from a race which had given her people so much trouble, overwhelmed me. I felt restored to me. *All* boundaries dissolved, then. There were all these dancing atoms and colors flowing, with shapes only dimly seen, and love was the only feeling that I had or knew. The washing continued but there was no blood, no flesh, only movement. My friend, the Navajo woman and I seemed to be not three persons but one, and I felt whole.

"The natural depth in man is the whole of creation. . . ."

The End: A Commencement

If this is truly "The End" of this book, then nothing has happened. If the reader carries it farther, adding to the experience of these pages, then something is happening.

There are two ways of carrying it farther. For some, this will be vertical—individuals going "farther" than any of us in this book have done—more deeply into themselves. The horizontal "farther" is many of us, individually, moving ever closer to being ourselves and letting others be themselves. This helps to release some persons to the vertical. Without the horizontal "farther," the vertical comes to nothing anyway. That's why what each one of us does is important, and equal to all others. This book is thus dedicated to all of us—to explore, express, and live more fully what it is to be human.

Be not like the child in *all* ways, child,
For the child in his impatient curiosity pries open the bud,
And the blossom he so wished to see
Is lost to everyone on earth forever.

Let thyself unfold thyself,
Without manipulation even by thee.
Let thy life also unfold itself around thee,
Without manipulation by thyself.

Removing impatience, let thy curiosity
Play lightly over this soft unfolding,
Which is essential to the flower of thyself,
For thyself and thy life and the world are one.

If you will but try this soft unfolding, child,
Then the flowering of thyself will be
Both more than you can possibly imagine,
And all that you have wished and longed to be.

About Bibliographies

Each of the professional papers in this book originally had a bibliography. I have omitted these except in the case of Van Dusen's paper which has not been published elsewhere. Anyone who wishes to go into this further can find the bibliographies in the journals, along with the complete papers which I have abridged. If anyone does not know how to follow through on this, any librarian can help him. In this way the person will learn not only how to find this particular material but also how to find a lot of other things he might want at other times. This seems to me better than for me to give more explicit instructions here for finding these particular papers.

I think it should be made possible for everyone to find what he wants, but that it should not be made so easy that he asks for much that he doesn't want, wouldn't use, and wouldn't do anything about otherwise. Advertisers act from a different base, but they're selling something: I'm not.

When Bernard de Voto wrote in Harper's Magazine that the National Parks were going to pot because of lack of money, that roads were in poor repair, and benches and tables and rest rooms were vandalized, I wrote to him, in my naivete, suggesting that this might be *good*. When I first visited the National Parks the roads were ruts, there were no benches or rest rooms, and there was no vandalism either. People who make that much effort to get somewhere, and put up with inconvenience, care enough for the place not to vandalize it. So, I suggested, let things run down and everything (in this respect) would take care of itself. Mr. de Voto replied that I was a "spiritual snob" and just the sort of person who was making things difficult for the Park Service.

I don't think so. People care for what they love, and if they don't love the National Parks they're not going to get much out of them anyway. It's just somewhere to go and be able to say I've been there. I think that anyone who really cares for pursuing the research and theory that I have omitted from some of these papers will find the papers themselves and use them. Then, they will not be wasted.

Some Related Reading

The following books and articles have been helpful to me in understanding myself and others, and the problem of being human. Some of them may seem out of place because they are in different categories but in fact they are all interrelated.

With regard to books, I have left out the usual information about publisher and in print or out of print, because the publishing world, too, is a changing world. Books that are out of print as I write may be in print by the time you read this. Books now in print may have been reversed too. Even publishers change, as a hardcover book put out by one publisher may appear in paperback by another. So any attempt to give "full" information can in fact turn out to be misleading—or *mis*information. I discovered this by looking up every title in this list.

In the case of articles, I include information about the journal because this stays put. The articles are still there, in the same journals.

These writings have been helpful to me. However I do not consider any of these books—or any book—necessary.

Albee, Edward: *The Zoo Story*. Play.
Armstrong, Charlotte: *A Dram of Poison*. Novel.
Axline, Virginia: *Dibs—In Search of Self*. Play therapy.
Bates, W. H.: *The Cure of Imperfect Sight by Treatment without Glasses.*
Beckett, Samuel: *Waiting for Godot*. Play.
Benoit, Hubert: *The Supreme Doctrine, Psychological Studies in Zen.*
Berne, Eric: *Games People Play: The Psychology of Human*

Relationships. Clarification of interpersonal interaction patterns. See especially the last section, "Beyond Games."

Bridgman, Percy W.: *The Way Things Are.* An individual assessment of the nature of experience and the world as man knows it. I have read the Preface and Introduction more than a dozen times, the rest of the book not at all.

Burrow, Trigant: *Preconscious Foundations of Human Experience.*

Cohen, Sidney: *The Beyond Within: The LSD Story.*

Eliot, T. S.: *The Cocktail Party.* Play.

Frankl, Viktor: *From Death Camp to Existentialism.*

Fromm, Erich: *The Art of Loving.*

Green, Hannah: *I Never Promised You a Rose Garden.* Novel.

Herrigel, Eugen: *Zen in the Art of Archery.*

Hesse, Hermann: *Siddartha.* Novel.

Holt, John: *How Children Fail.*

Huxley, Aldous: *The Art of Seeing.* The Bates method interpreted by Huxley. (See also *Bates.*) *Doors of Perception.* Mescaline experience.

I Ching: *The Book of Changes.* I depart from my own rule here because the 2-volume Pantheon edition, translated from the Chinese by Richard Wilhelm, into English by Carey Baynes, seems to me so much superior to other editions which I have seen.

Jackson, Shirley: *The Bird's Nest.* (hardcover title) Novel. *Lizzie.* (paperback title of the same book).

Klee, James B.: "The Absolute and the Relative." *Darshana International,* Moradabad, India, vol. 4, nos. 1 and 2, 1964.

Krishnamurti, J.: *Think on These Things. Education and the Significance of Life. Life Ahead. The First and Last Freedom.* Psychology/philosophy.

Kubie, Lawrence S.: *Neurotic Distortion of the Creative Process.*

Marc, Franz: *Watercolors, Drawings, Writings.* Ed. by Klaus Lankheit.

Maslow, A. H.: *Motivation and Personality. Toward a Psychology of Being.*

McKellar, Peter: *Imagination and Thinking.*

Morgan, Charles: *The Voyage.* Novel.

Moustakas, Clark: *The Self: Explorations in Personal Growth. Creativity and Conformity.*

Ortega y Gasset, Jose: *The Modern Theme*. Philosophy.
Ratner, Herbert: *Medicine*. A booklet published by the Center for the Study of Democratic Institutions, Box 4068, Santa Barbara, Calif.
Rogers, Carl R.: *Client-centered Therapy*. *On Becoming a Person*.
Schachtel, Ernest G.: *Metamorphosis: On the Development of Emotion, Perception, Attention and Memory in the Child*. (As they relate to creativity and psychological health. See especially the chapter "Memory and Childhood Amnesia.")
Stevens, John O.: "Determinism; prerequisite for a meaningful freedom." *Review of Existential Psychology and Psychiatry*, Fall, 1967, Vol. VII. No. 3.
Sturgeon, Theodore: *More Than Human*. Novel. "The [Widget], the [Wadget], and Boff." A story, in *Aliens 4*.
Suzuki, D. T.: *Studies in Zen*.
Szasz, Thomas: *Law, Liberty and Psychiatry*.
Tauber (Edward S.) and Green (Maurice R.): *Prelogical Experience: An Inquiry into Dreams and Other Creative Processes*.
Tolstoy, Leo: *Twenty-three Tales*. The following stories: "Three Questions," "What Men Live By," "The Three Hermits," "The Story of Ivan the Fool," "The Godson," "Two Old Men."
Weaver, Warren: "The Imperfections of Science." *American Scientist*, vol. 45, no. 1, March 1961.
Williams, Roger: *Biochemical Individuality*.

About the Authors

Carl R. Rogers

Resident Fellow, Western Behavioral Sciences Institute, La Jolla, California. Formerly, professor of psychology and psychiatry at the University of Wisconsin; professor of psychology and organizer and head of the Counseling Center, University of Chicago, among other professional posts. Past president of the American Association for Applied Psychology, the American Psychological Association, the American Academy of Psychotherapists. His most recent book is *On Becoming a Person* (Houghton Mifflin, 1961).

"I fall *far* short of achieving real communication—person-to-person—all the time, but moving in this direction makes life for me a warm, exciting, upsetting, troubling, satisfying, enriching, and above all a worthwhile venture."

Eugene T. Gendlin

Assistant Professor, Departments of Philosophy and Psychology, University of Chicago. Both his M.A. and his Ph.D. were in Philosophy, University of Chicago. Editor, *Psychotherapy: Theory, Research and Practice* (Journal). Author, *Experiencing and the Creation of Meaning* (Free Press [Macmillan], New York, 1962). His collected papers have been translated by T. Murase, Tokyo, and published (in Japan by Maki) 1966.

"I will not let you (or me) make me dishonest, insincere, emotionally tied up or constricted, or artificially nice and social, if I can help it."

John M. Shlien

Formerly Professor of Psychology and Human Development at the University of Chicago; now Professor of Counseling Psychology at Harvard University in the Graduate School of Education. Editor, Volume III, *Research in Psychotherapy;* contributor to many books and journals.

"What's to say? I have the feeling that everybody knows everything so far as human interaction goes, and that we only choose to ignore or to forget. That's why social science is so hard to teach. The people I love best are the ones who dig daffodils as well as ancient history, who put together idealism and intelligence with rare force and gentleness."

Wilson Van Dusen

Chief Psychologist, Mendocino State Hospital, Talmage, California. Professor, Sonoma State College, California. Ph.D. University of Ottawa, Canada. Thesis involved an extension of Einstein's 4-space into higher dimensions, showing that 5- to 7-space has properties of mind. Besides electronics and theology, his major interest is in understanding and describing inner states of man. His most recent work —on religious experiences under LSD, and capturing psychotics' experiences of hallucinations—is in accord with his special bent towards areas of human experience that are widely overlooked.

"I enjoy looking closely at the very nature and quality of human experience—the familiar, near at hand, ever present, answer to its own questions, implying all because it contains all, ground of all feeling, clearing house of all understanding, simple beyond knowing, our very lives."

Barry Stevens

High School drop-out, 1918, because what she wanted to know, she couldn't learn in school.

www.ingramcontent.com/pod-product-compliance
Ingram Content Group UK Ltd.
Pitfield, Milton Keynes, MK11 3LW, UK
UKHW021414140225
4604UKWH00003B/39